PE

principles and practice of engineering

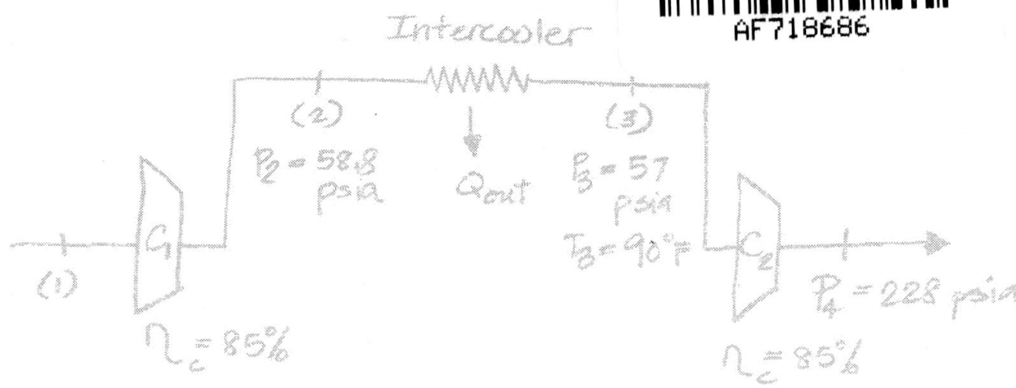

NCEES
advancing licensure for engineers and surveyors

mechanical: THERMAL and FLUIDS SYSTEMS

sample questions + solutions

280 Seneca Creek Road
Clemson, SC 29631
800-250-3196
www.ncees.org

ISBN 978-1-932613-61-2

Printed in the United States of America
November 2010

Paper used in the printing of NCEES study materials meets environmental guidelines of the Rainforest Alliance, Elemental Chlorine Free (ECF) papers, and Sustainable Forestry Initiative (SFI).

CONTENTS

About the publisher

NCEES is a national nonprofit organization dedicated to advancing professional licensure for engineers and surveyors. It develops, administers, and scores the examinations used for engineering and surveying licensure in the United States. It also facilitates professional mobility and promotes uniformity of the U.S. licensure processes through services for its member licensing boards and licensees. These services include the records program, study materials, credentials evaluations, exam administration, and more.

NCEES is composed of member licensing boards from throughout the United States and its territories.

Updates on exam content and procedures

NCEES.org is our home on the Web. Visit us there for updates on everything exam-related, including specifications, exam-day policies, scoring, and corrections to published study materials. This is also where you will register for the exam and find the additional steps you should follow in your state to be approved for the exam.

Exam-day schedule

Be sure to arrive at the exam site on time. Late-arriving examinees will not be allowed into the exam room once the proctor has begun to read the exam script. The report time for the exam will be printed on your Exam Authorization. Normally, you will be given 1 hour between morning and afternoon sessions.

Admission to the exam site

To be admitted to the exam, you must bring two items: (1) your Exam Authorization and (2) a current, signed, government-issued photo identification.

Candidate Agreement

The NCEES Candidate Agreement describes references and other personal materials allowed in the exam room. It also describes other exam policies designed to protect exam content. This document will be printed in your exam book. It is also available for download at NCEES.org. If you register to take an NCEES exam, you are strongly encouraged to review this document prior to exam day, as you will be asked to affirm that you abide by the policies and procedures it describes.

Scoring and reporting

NCEES typically releases exam results to its member licensing boards 8–10 weeks after the exam. Depending on your state, you will be notified of your exam result online through your My NCEES account or via postal mail from your state licensing board. Detailed information on the scoring process can be found at NCEES.org.

EXAM SPECIFICATIONS

NCEES Principles and Practice of Engineering Examination
MECHANICAL BREADTH and THERMAL and FLUIDS SYSTEMS Depth
Exam Specifications

Effective Beginning with the October 2008 Examinations

- The mechanical engineering examination is a breadth and depth examination. This means that examinees work the breadth (AM) exam and **one** of the three depth (PM) exams. The AM questions are the same for all Mechanical PM modules; that is, regardless of the PM module an examinee chooses, the AM questions will be the same in each exam.
- The three areas covered in the mechanical engineering examination are HVAC and Refrigeration; Mechanical Systems and Materials; and Thermal and Fluids Systems. The breadth exam contains questions from these three areas of mechanical engineering. The depth exams focus more closely on a single area of practice in mechanical engineering.
- Examinees work all questions in the morning session and all questions in the afternoon module they have chosen. Depth results are combined with breadth results for final score.
- The exam is an 8-hour open-book exam. It contains 40 multiple-choice questions in the 4-hour AM session, and 40 multiple-choice questions in the 4-hour PM session.
- Questions in Sections I–IV of the breath module will be in *either* USCS or SI units. Questions in Section V of the breadth module will be in USCS units. Questions in the depth module will be in USCS units.
- The exam is developed with questions that will require a variety of approaches and methodologies including design, analysis, and application. Some questions may require knowledge of engineering economics.
- The knowledge areas specified as examples of kinds of knowledge are not exclusive or exhaustive categories.
- The specifications for the **AM exam** and the **Thermal and Fluids Systems PM exam** are included here.

MECHANICAL BREADTH Exam Specifications

	Approximate Percentage of AM Exam
I. Basic Engineering Practice	**30%**
A. Basic Engineering Practice	30%
1. Engineering terms and symbols	
2. Economic analysis	
3. Project management	
4. Interpretation of technical drawings	
5. Electrical concepts	
6. Units and conversions	
II. Mechanical Systems and Materials	**20%**
A. Principles	13%
1. Statics and dynamics	
2. Strength of materials	
3. Stress analysis	
4. Fatigue theory	
B. Applications	7%
1. Mechanical components (e.g., springs, gears, pressure vessels)	
2. Joints and fasteners (e.g., welding, bolts, adhesives)	
3. Vibration/dynamic analysis	
4. Materials selection (e.g., corrosion, weight, strength)	
III. Hydraulics and Fluids	**17%**
A. Principles	7%
1. Compressible flow	
2. Incompressible flow	
B. Applications	10%
1. Hydraulic and fluid equipment (e.g., pumps, turbines, compressors)	
2. Piping systems and components	
IV. Energy/Power Systems	**15%**
A. Principles	7%
1. Thermodynamic cycles	
2. Thermodynamic properties	
3. Energy balances	
4. Mass balances	
5. Heat transfer	
6. Combustion	
B. Applications	8%
1. Power conversion systems	
2. Energy/power equipment (e.g., turbines, boilers, engines)	
3. Heat exchangers	
V. HVAC/Refrigeration	**18%**
A. Principles	10%
1. Psychrometrics	
2. Refrigeration cycles	
3. Heat transfer	
B. Applications	8%
1. HVAC/refrigeration systems	
2. HVAC/refrigeration components (e.g., air handlers, compressors)	
3. Heating/cooling loads	

MECHANICAL–THERMAL and FLUIDS SYSTEMS Depth Exam Specifications

	Approximate Percentage of PM Exam
I. Principles	**45%**
A. Materials Properties (e.g., density, viscosity)	5%
B. Fluid Mechanics	10%
1. Compressible fluids (e.g., Mach number, nozzles, diffusers)	
2. Incompressible fluids (e.g., friction factor, Reynolds number, lift, drag)	
C. Heat Transfer Principles (e.g., convection, conduction, radiation)	10%
D. Mass Balance Principles (e.g., evaporation, dehumidification, combustion)	7%
E. Thermodynamics	10%
1. Thermodynamic cycles (e.g., combined, Brayton, Rankine)	
2. Thermodynamic properties (e.g., enthalpy, entropy)	
3. Energy balances (e.g., 1st and 2nd Laws)	
4. Combustion (e.g., stoichiometrics)	
F. Related Principles	3%
1. Strength of materials (e.g., stress-strain, yield strength)	
2. Fatigue theory (e.g., Goodman diagram)	
3. Statics and dynamics	
4. Stress analysis (e.g., pipe stress, pipe hangers, hoop stress)	
5. Psychrometrics (e.g., dew point, relative humidity)	
6. Welding (e.g., processes, symbols)	
7. Safety (e.g., OSHA, industrial, ergonomics, sanitation)	
8. Quality control/quality assurance	
II. Applications	**55%**
A. Equipment	18%
1. Pumps	
2. Turbines	
3. Compressors, fans, blowers	
4. Boilers, steam generators	
5. Engines and drive trains	
6. Pressure vessels	
7. Heat exchangers/condensers/feed water heaters	
8. Cooling towers	
9. Control devices (valves, flow measurement)	
B. Systems	32%
1. Power hydraulics	
2. Pneumatic	
3. Fluid distribution	
4. Power conversion	
5. Energy recovery	
6. Cooling/heating cycles	
7. Power cycles	
C. Codes and Standards	5%

MECHANICAL
MORNING SAMPLE QUESTIONS

A psychrometric chart (normal temperature at sea level) is provided on page 89 for your possible use. Unless stated otherwise, assume sea level conditions.

MECHANICAL MORNING SAMPLE QUESTIONS

101. Poisson's ratio is a term that arises in:

(A) heat transfer
(B) HVAC
(C) mechanical behavior
(D) vapor power cycles

102. Which of the following is used as a measure of energy efficiency of refrigeration systems?

(A) COP
(B) Hydraulic head
(C) Latent heat
(D) SHR

103. A system with an initial cost of $90,000 has an operating cost of $2.50/hr and is operated 16 hr/day for 300 days/year. After 10 years, the system has a salvage value of $10,000. Assuming the interest rate is constant at 10%, the total annualized cost over the 10-year life is most nearly:

(A) $12,000
(B) $14,000
(C) $20,000
(D) $26,000

104. You manage a manufacturing production department that operates 80 hr/week. The process has a production rate of 60 units/hr, with a scrap rate of 1 out of 120 units. With a 10% lost production time allowance, the number of good units that can be made per week is most nearly:

(A) 4,284
(B) 4,320
(C) 4,760
(D) 4,800

105. A project consists of assembling a system using parts from Source X and Source Y.

Task	Description	Follows Task	Duration (days)
A	Procure parts from Source X	None	2
B	Procure parts from Source Y	A	2
C	Delivery of parts from Source X	B	7
D	Delivery of parts from Source Y	A	11
E	Assemble parts and complete project	C and D	2

If the manufacturing of parts from Source X is delayed by 3 days, now requiring 10 days to obtain those parts, the total completion delay (days) will be most nearly:

(A) 0
(B) 1
(C) 2
(D) 3

106. A commercial product is sold in 55-gal drums. It is made up primarily of inert ingredients, to which an active ingredient is added. The active ingredient can be purchased in two different formulations as shown.

Formulation	Cost/lb	Active Ingredient (units/lb)
1	$8	100
2	$10	200

The amount of Formulation 1 used per drum cannot exceed 3 lb, and the amount of Formulation 2 used per drum cannot exceed 2 lb. The total cost for the active ingredient cannot exceed $36/drum. The largest number of units of active ingredient that can be added per drum under these constraints is most nearly:

(A) 400
(B) 540
(C) 600
(D) 700

107. A machining operation creates a hole for a 3/8-16NC × 0.75-in.-long countersunk hex head screw, as shown in **Figure 1**. A cutter is used to create the countersunk recess as shown in **Figure 2**. The process steps are as follows:

1. Drill through with a tap drill diameter of 5/16 in.
2. Countersink the tap drill hole to a depth of Dimension A, so that the screw head will be flush with the part surface.
3. Tap the hole 3/8-16NC.

To complete Step 2, the necessary depth (in.) for Dimension A is most nearly:

(A) 0.202
(B) 0.232
(C) 0.234
(D) 0.269

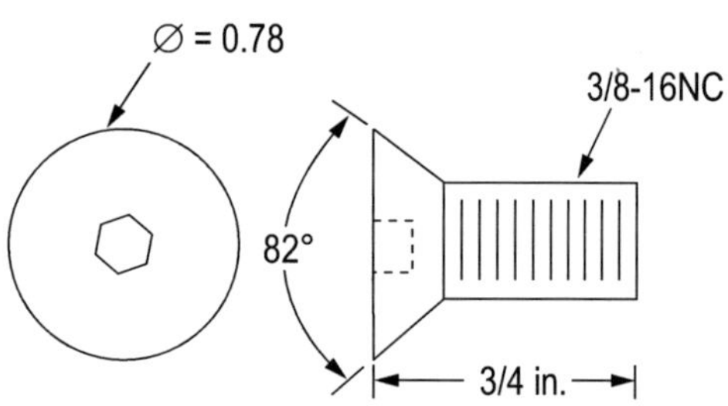

FIGURE 1

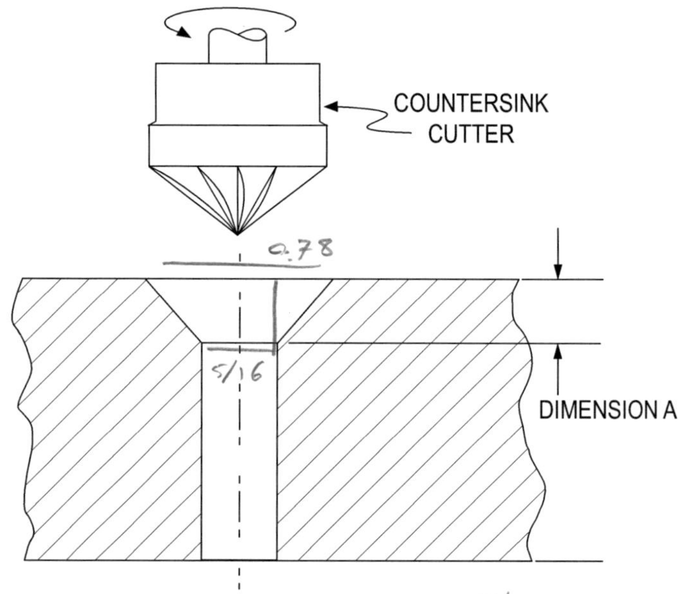

FIGURE 2

108. The minimum distance between the surfaces of Holes A and B is most nearly:

(A) 0.20
(B) 0.30
(C) 0.40
(D) 0.50

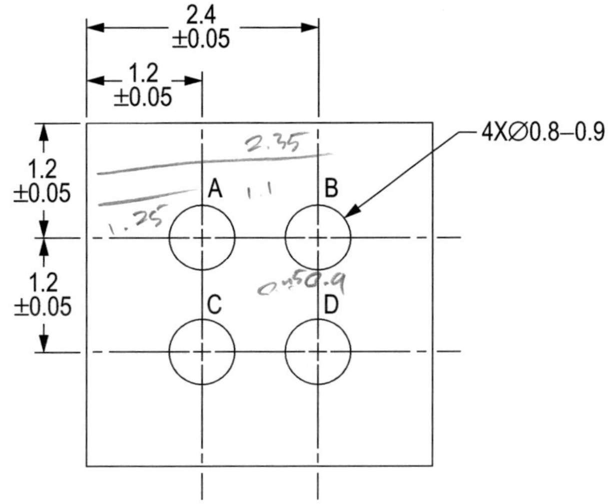

109. Given the dc electrical circuit shown, which of the following statements is true?

(A) $V_1 + V_2 = IR_1 + IR_2$

(B) $\dfrac{V_1}{R_1 + R_2} = \dfrac{V_2}{R_3}$

(C) Equivalent resistance for the loop, $R_e = R_1 + R_2 + R_3$

(D) $\dfrac{1}{R_1} + \dfrac{1}{R_2} = \dfrac{1}{R_3}$

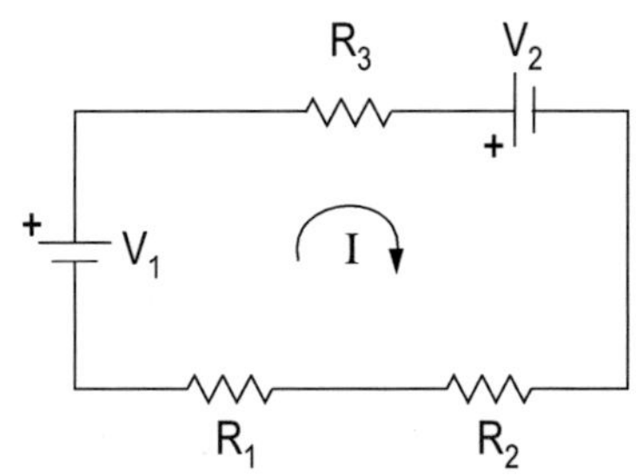

110. A 230-V_{rms}, single-phase, 10-hp motor operates at 80% power factor and its rated load. If the efficiency of the motor is 65%, the current (amperes) drawn is most nearly:

(A) 12
(B) 17
(C) 40
(D) 62

111. Water flows over Niagara Falls at an average rate of 4,000,000 cfm and drops 170 ft to the base of the falls. The rate of change of gravitational potential energy (MW) is most nearly:

(A) 400
(B) 700
(C) 1,000
(D) 57,000

112. Fracture toughness of a material is measured in units of ksi$\sqrt{\text{in.}}$ in the English system of units. A material with a fracture toughness of 35 ksi$\sqrt{\text{in.}}$ will have a fracture toughness in units of MPa$\sqrt{\text{m}}$ that is most nearly:

(A) 1
(B) 6
(C) 32
(D) 38

113. A 1/2-in.-diameter pipe hanger rod is loaded in tension with a 2,300-lb force. If the rod is 6 in. long and is made of Material A, the total elongation (in.) it will experience is most nearly:

(A) 0.0004
(B) 0.002
(C) 0.005
(D) 0.01

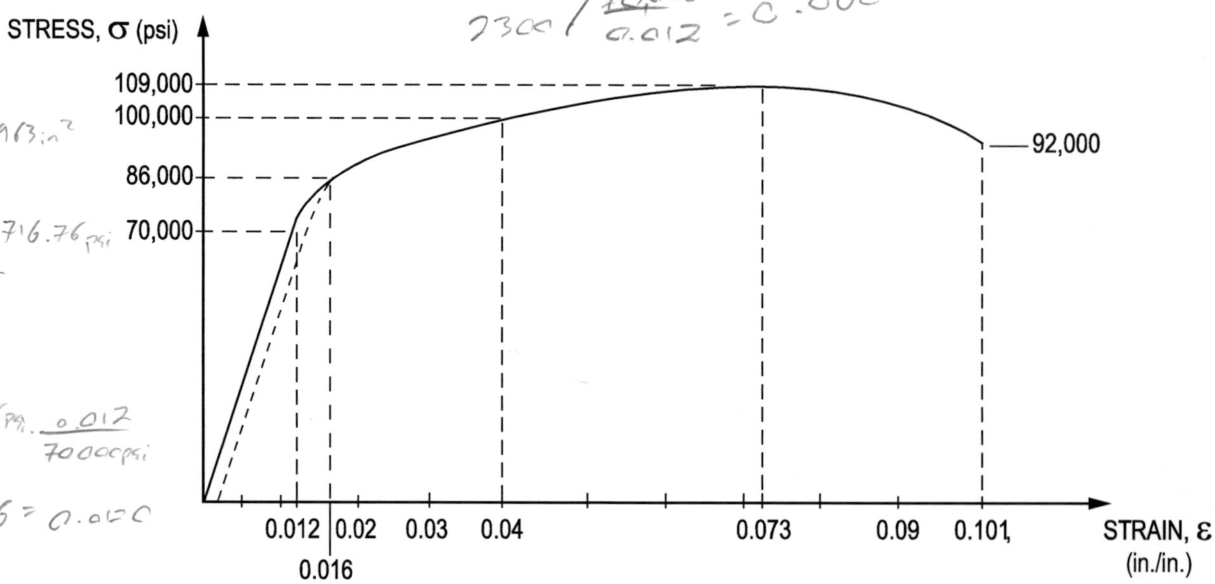

MATERIAL A

114. A cylindrical pressure vessel with hemispherical end caps is unrestrained. The diameter is 1 m, the wall thickness is 1 cm, and the internal gage pressure is 3 MPa. If the tensile yield strength S_{YT} is 450 MPa and the von Mises failure criterion is used, then the factor of safety $(S_{YT/\sigma effective})$ is most nearly:

(A) 1.5
(B) 1.7
(C) 3.0
(D) 3.5

115. A cord is wrapped halfway around a 2-in.-O.D. pipe as shown. The coefficient of friction (dynamic and static) between the cord and pipe is μ = 0.35. A box with a weight of 50 lb is suspended from one end of the cord and a force P applied to the other end. The minimum force, P (lb), required to raise the 50-lb box is most nearly:

(A) 16.7
(B) 50.0
(C) 85.0
(D) 150.0

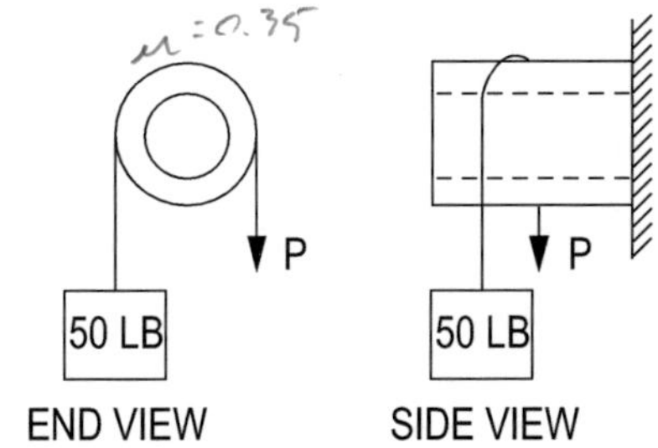

116. A linear spring-mass system consisting of a spring with a spring constant of 20 lbf/in. and a mass of 10 lbm undergoes free vibration. The linear frequency (cycles/sec) of this vibration is most nearly:

(A) 0.78
(B) 1.3
(C) 4.4
(D) 28

117. Material compatibility concerns require that a steel spring made from wire with a shear modulus of 11×10^6 psi be redesigned. Brass with a shear modulus of 5.5 × 106 psi will be used instead. The steel wire has a diameter of 0.15 in. The spring diameter and the number of turns must be the same for the steel and brass springs. If the spring constant of the brass spring is to be the same as that of the steel spring, the diameter (in.) of the brass wire must be most nearly:

(A) 0.12
(B) 0.15
(C) 0.18
(D) 0.30

118. A metal is needed consisting of nearly equal amounts of pearlite and ferrite when cooled at a slow rate from 1,800°F to room temperature. The metal that best meets this need is:

(A) Class 40 grey cast iron

(B) AISI 304

(C) AISI 1040

(D) AISI 1095

IRON-CARBON PHASE DIAGRAM

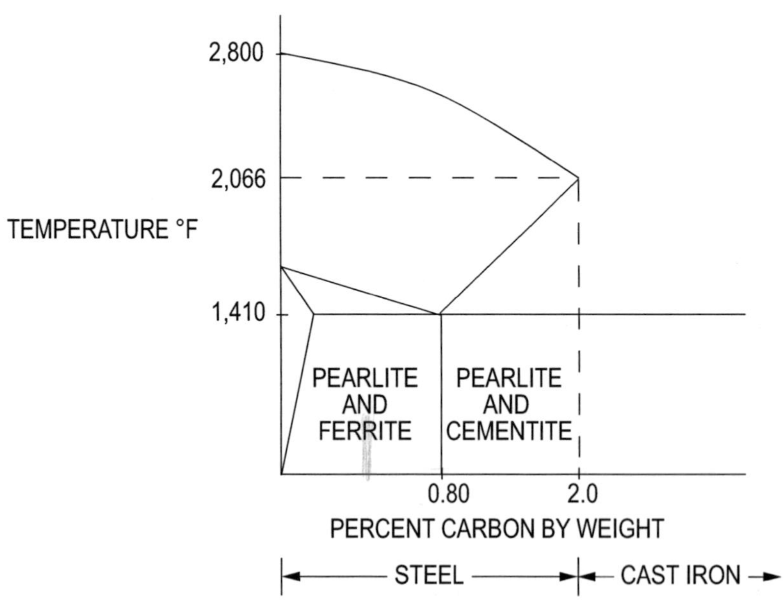

119. Which of the following factors is **least** likely to affect the fatigue life of a steel spring?

(A) Alternating stresses
(B) Cycle period
(C) Primary stresses
(D) Surface roughness

120. The 2,500-lbf load shown in the figure is supported by a beam that is secured by six 3/4-in.-diameter bolts in a hexagonal array with neighboring bolts 3 in. apart. The shear stress is greatest in which bolt?

(A) 1
(B) 2
(C) 3
(D) 4

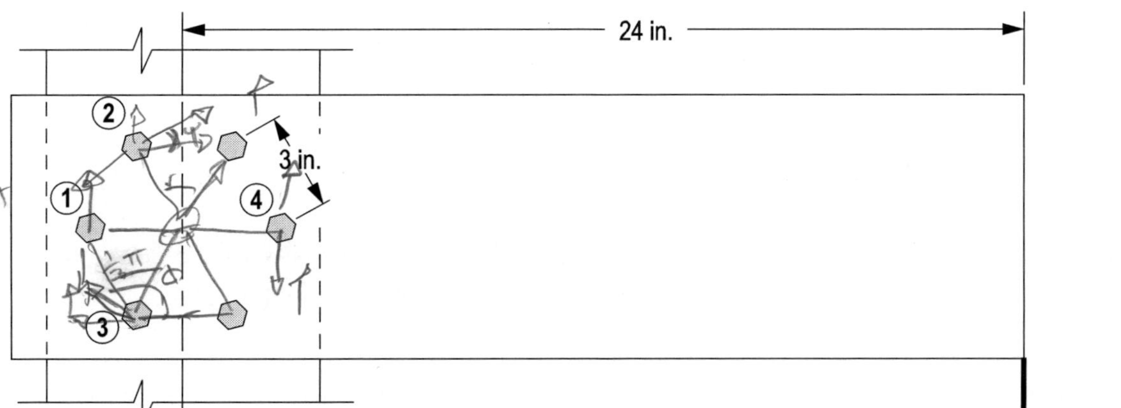

MECHANICAL MORNING SAMPLE QUESTIONS

121. The Mach number just upstream of a normal shock wave in air (k = 1.4) is 4.5. The Mach number downstream of the normal shock wave is most nearly:

(A) 0.42
(B) 1.00
(C) 2.00
(D) 4.50

122. The figure below shows water that flows in from two pipes and exits from another pipe. The velocity of flow (fps) in the 12-in-I.D. pipe is most nearly:

(A) 1.9
(B) 2.4
(C) 3.1
(D) Cannot be determined from information given

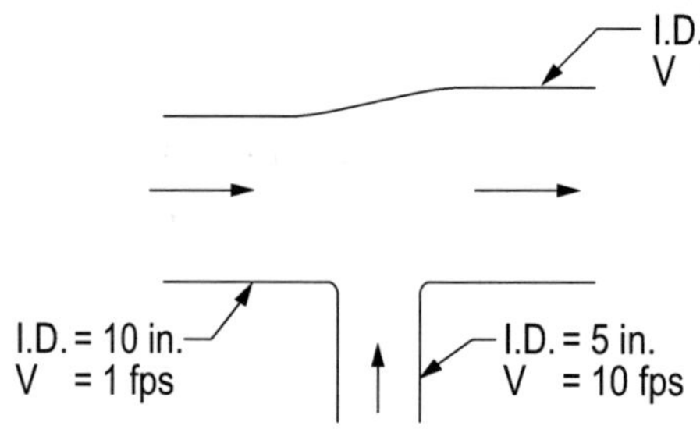

123. A fluid with a specific gravity of 0.8 and a kinematic viscosity of 1.5 centistokes (1 centistoke = 1.0764×10^{-5} ft^2/sec) flows through a 4-in.-I.D. hose at 5 fps. The Reynolds number for this flow is most nearly:

(A) 1.2×10^6
(B) 1.0×10^6
(C) 1.0×10^5
(D) 8.3×10^4

124. A centrifugal pump operating at 1,750 rpm consumes 1,000 hp. If the speed of the pump is reduced to 1,000 rpm, the power consumption (kW) will be most nearly:

(A) 426
(B) 243
(C) 187
(D) 139

125. A centrifugal pump is sized to deliver 200 gpm of water at 60°F against a head of 60 ft of water. It is operating at 1,750 rpm. The pump efficiency is 0.75, and the motor efficiency is 0.83. If the electricity cost is \$0.06/kWh, the electrical cost to operate the pump 8,000 hr/yr is most nearly:

(A) \$984
(B) \$1,086
(C) \$1,200
(D) \$1,750

126. Water at 60°F flows through a 1-in brass pipe (1.063 in. I.D.) at 5 fps. The Moody friction factor is 0.025. The pressure drop (psi) over a 100-ft length of this pipe is most nearly:

(A) 0.40
(B) 4.8
(C) 13
(D) 150

127. Twenty gpm of 60°F water enters the parallel piping system shown in the figure. Branch A consists of Sch. 40 1-in. steel pipe with an equivalent length of 80 ft. Branch B consists of 100 ft of Sch. 40 1-in. steel pipe, four 90° elbows with equivalent lengths of 5.2 ft each, and a globe valve with an equivalent length of 29 ft. The flow rate (gpm) through Branch B is most nearly:

(A) 7
(B) 8
(C) 10
(D) 13

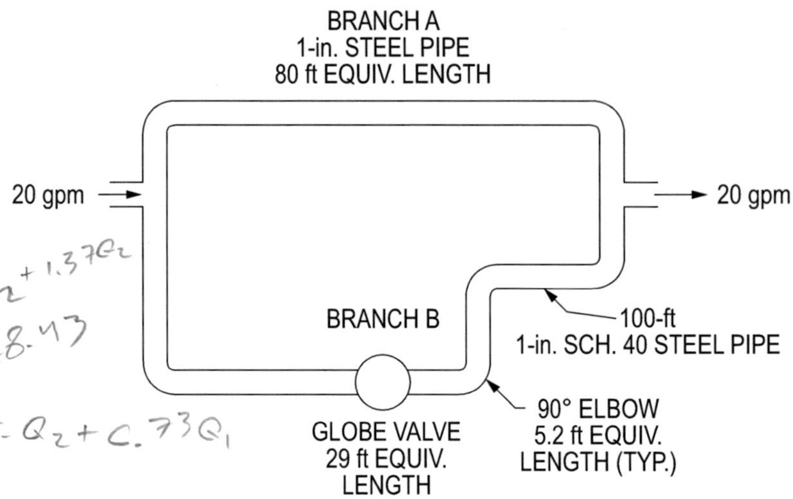

128. A Carnot cycle operates between 400°F and 250°F. The maximum theoretical efficiency of the cycle is most nearly:

(A) 0.17
(B) 0.21
(C) 0.38
(D) 0.60

129. Water (100 lbm/min, 60°F) and steam (100 lbm/min, 600°F), both at atmospheric pressure, enter an insulated chamber through separate inlets. After thorough mixing, 200 lbm/min of the product is withdrawn at atmospheric pressure. The quality of the steam in the product stream is most nearly:

(A) 0.12
(B) 0.43
(C) 0.52
(D) 0.60

130. The expression for the sensitivity of a pressure thermometer (change in pressure per unit change in temperature of an ideal gas) is:

(A) $\frac{V}{n\overline{R}}$

(B) $\frac{nV}{\overline{R}}$

(C) $\frac{n}{\overline{R}V}$

(D) $\frac{n\overline{R}}{V}$

131. Consider the following stoichiometric reaction:

$$C_9H_{18} + 13.5\ O_2 + 50.76\ N_2 \rightarrow 9\ CO_2 + 9\ H_2O + 50.76\ N_2$$

The mass fraction of water in the flue gas is most nearly:

(A) 5.8%
(B) 8.2%
(C) 10.7%
(D) 13.1%

132. Steam (600°F, 300 psia) is expanded in an 80% adiabatic efficiency turbine to 3 psia. The enthalpy (Btu/lbm) of the resulting 3-psia steam is most nearly:

(A) 880
(B) 970
(C) 1,040
(D) 1,120

133. The cooling medium in a counterflow heat exchanger increases in temperature from 60°F to 110°F, while the process solution being cooled decreases from 150°F to 130°F. The logarithmic mean temperature difference (LMTD) for this counterflow heat exchanger is most nearly:

(A) 33°F
(B) 47°F
(C) 54°F
(D) 70°F

134. Based on the psychrometric chart provided on page 89, sea level air at a dry-bulb temperature of 95°F and 25% rh has a dew point of most nearly:

(A) 38°F
(B) 54°F
(C) 68°F
(D) 82°F

135. The partial pressure (psi) of water in sea level air at 80°F and 60% rh is most nearly:

(A) 0.0135
(B) 0.30
(C) 0.35
(D) 0.50

136. An automotive air-conditioning system based on R-134a operates between 250 psia and 40 psia. A pressure-enthalpy diagram for R-134a is presented on the opposite page. If 250-psia saturated liquid R-134a is throttled to 40 psia, the change [Btu/(lbm-°R)] in entropy ($S_{FINAL} - S_{INITIAL}$) will be most nearly:

(A) –0.07
(B) 0.01
(C) 0.11
(D) 0.15

136. (Continued)

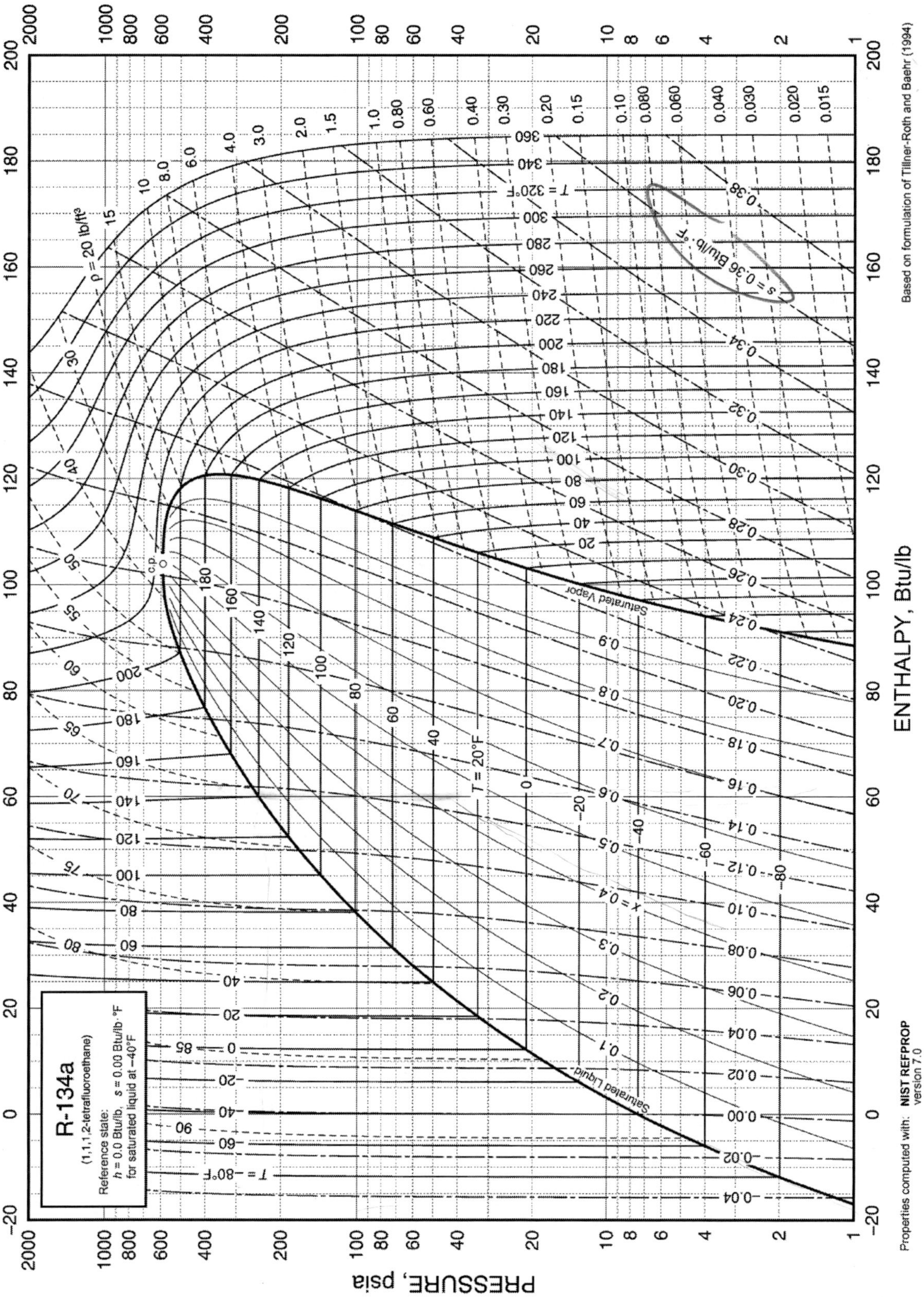

Fig. 8 Pressure-Enthalpy Diagram for Refrigerant 134a

2005 ASHRAE Handbook–Fundamentals, I-P edition, p. 20.16, Figure 8, © American Society of Heating, Refrigerating, and Air-Conditioning Engineers, Inc., www.ashrae.org.

GO ON TO THE NEXT PAGE

137. An expansion valve would be used in which leg of the refrigeration cycle shown in the figure?

(A) A–B

(B) B–C

(C) C–D

(D) D–A

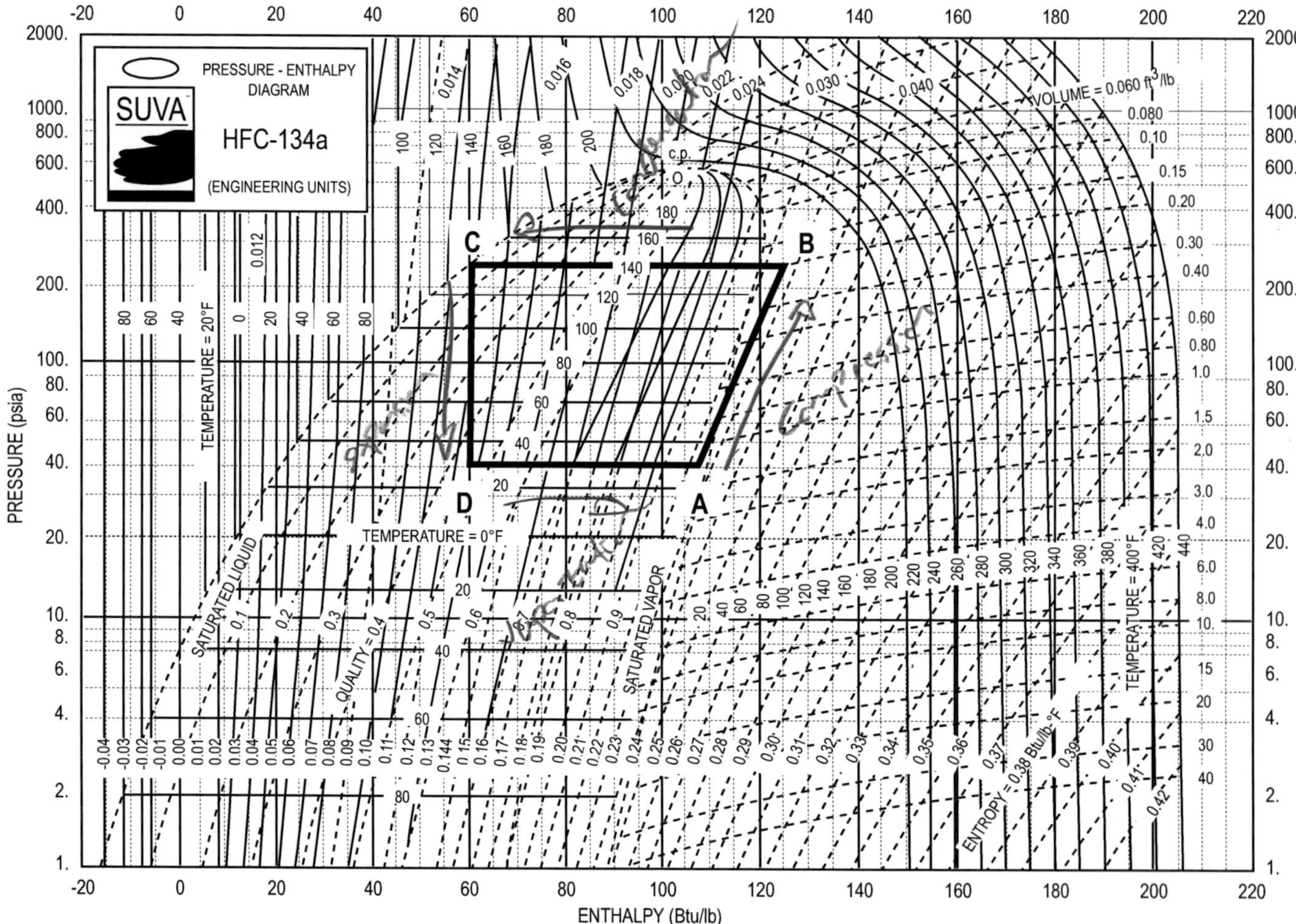

From DuPont Refrigerants

138. A room contains equipment that consumes 3,000 W of power. In addition, 5 lb/min of water vapor is released into the room. The resulting cooling load (Btu/hr) is most nearly:

(A) 3,000
(B) 4,500
(C) 10,000
(D) 15,000

139. A refrigeration system with a capacity of 1 refrigeration ton has a compressor work of 1 kW. The coefficient of performance of the system is most nearly:

(A) 0.3
(B) 1
(C) 3.5
(D) 12

140. A counterflow heat exchanger with a surface area of 20 ft^2 condenses saturated H-134a vapor at 250 psia (T = 140°F, h = 120 Btu/lbm) to saturated liquid at 250 psia (h = 60 Btu/lbm). The inlet temperature of the cooling water is 60°F, and the outlet temperature of the cooling water is 75°F. The overall heat-transfer coefficient for the heat exchanger is 8 Btu/(hr-ft^2-°F). The rate (lbm/hr) at which the H-134a is condensed is most nearly:

(A) 20
(B) 40
(C) 100
(D) 200

This completes the morning session. Solutions begin on page 51.

THERMAL AND FLUIDS SYSTEMS
AFTERNOON SAMPLE QUESTIONS

A psychrometric chart (normal temperature at sea level) is provided on page 89 for your possible use. Unless stated otherwise, assume sea level conditions.

501. A U-tube manometer containing an unknown fluid is connected to a tank containing a gas. The tank pressure is 19.4 psia, and atmospheric pressure is 14.7 psia. If the fluid height in the manometer is 35.0 in., the density (lbm/ft^3) of the manometer fluid is most nearly:

(A) 1.6
(B) 19.3
(C) 62.4
(D) 232

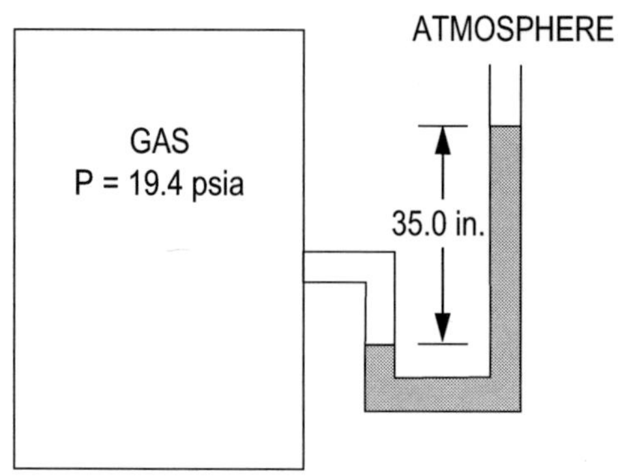

502. Steam at 900,000 lbm/hr exits a low-pressure steam turbine at 2.0 psia with 7% moisture content and a velocity of 300 fps. The required cross-sectional area (ft^2) is most nearly:

(A) 56.3
(B) 135
(C) 145
(D) 156

503. An airplane is flying at a Mach number of 0.85. If the absolute ambient pressure is 20.64 kPa and the ambient temperature is –56.7°C, the airplane velocity (m/s) is most nearly:

(A) 251
(B) 292
(C) 344
(D) 395

504. Air with the following properties enters the compressor of an aircraft gas turbine.

P = 5.45 psia
T = 430°R
V = 600 fps
$\dot{m}$ = 150 lbm/sec

The cross-sectional area (ft^2) at the inlet is most nearly:

(A) 1.64
(B) 2.70
(C) 3.27
(D) 7.30

505. The flow of water in a 10-in. Schedule 40 pipe is to be metered. The temperature of the water is 100°F, and the static pressure upstream of the meter is 20 psig. The density of the water is 62 lbm/ft^3. Assume the flowmeter is a square-edged orifice with a diameter ratio of 0.5. When the flow rate is 1,200 gpm and the flow coefficient is 0.5, the pressure drop across the orifice (lbf/in^2) is most nearly:

(A) 2.6
(B) 6.6
(C) 8.4
(D) 10.3

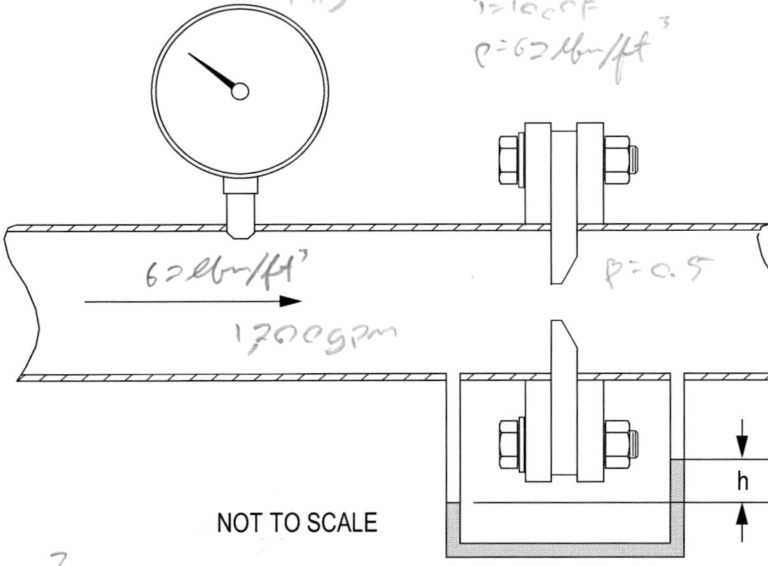

506. A motor-driven pump draws water from a large reservoir as shown in the figure below. The water level in the reservoir remains constant. The pump is required to move 300 gpm. The suction pipe is 4-in. Schedule 40 black steel, and the discharge pipe is 3-in. Schedule 40 black steel. The large reservoir contains water at 193°F and 10 psia. The pressure at the pump discharge must be 120 psig.

Water properties at 193°F:

Density, ρ	60.3 lbm/ft^3
Kinematic viscosity, υ	0.356×10^{-5} ft^2/sec
Vapor pressure	23.06 ft of fluid

Pipe sizes:
- 3 in. nominal is 3.068 in. I.D.
- 4 in. nominal is 4.026 in. I.D.

With a friction factor of 0.018, the line losses (ft) between the reservoir and pump, including entrance losses and friction losses through pipe, are most nearly (entrance K = 0.5, elbow K = 0.9):

(A) 11.2
(B) 5.6
(C) 2.8
(D) 1.4

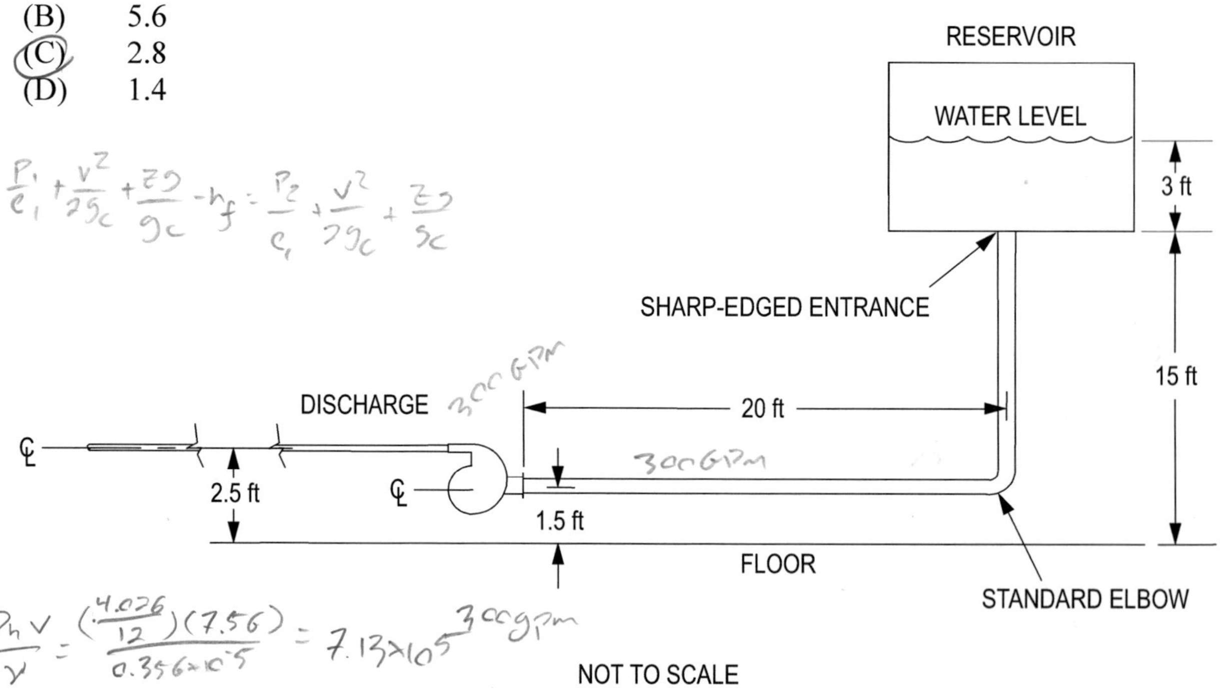

507. A solar-thermal power system pumps and stores its hot (1,000°F) liquid sodium in an insulated overhead storage tank, as shown in the figure below. The pump delivers sodium at the rate of 50 lbm/sec.

Use the following properties of sodium at 1,000°F:

Density	51.4 lbm/ft^3
Specific heat	0.301 Btu/(lbm-°F)
Thermal conductivity	37.6 Btu/(hr-ft-°F)
Dynamic viscosity	0.000156 lbm/(ft-sec)

Use the following data for the sodium pipe:

Nominal size	4-in Schedule 40 stainless steel
Inside diameter	4.026 in.
Darcy friction factor	0.01

The coefficient [Btu/(hr-ft^2-°F)] for convective heat transfer at the inside surface of the pipe is most nearly:

(A) 275
(B) 1,455
(C) 3,320
(D) 28,150

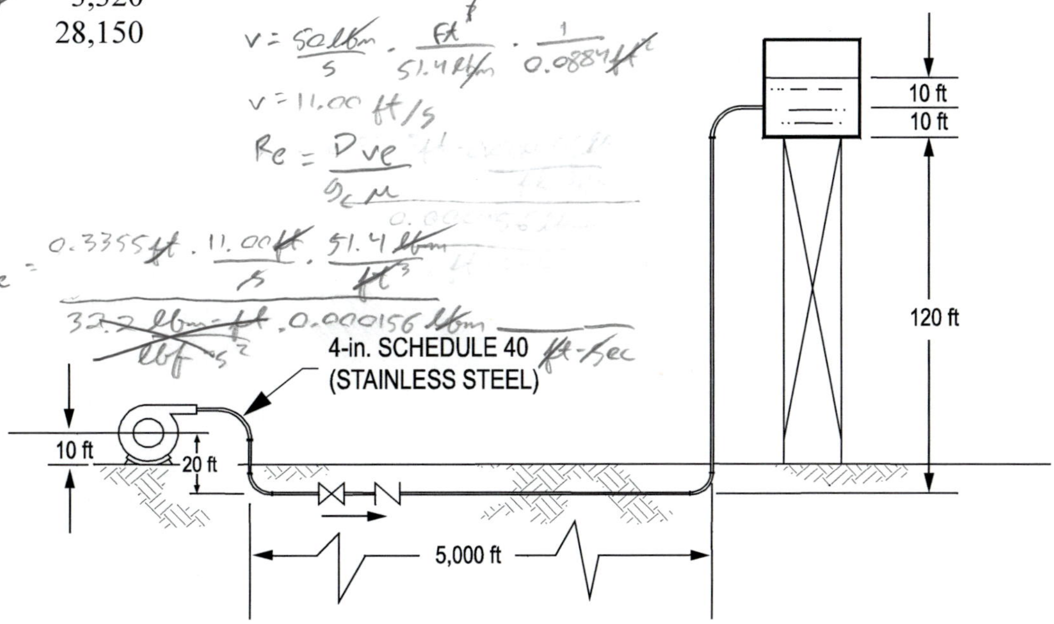

NOT TO SCALE

508. In a shell-and-tube heat exchanger, 10,000 lbm/hr of water is heated from 40°F to 160°F. The heating medium is 37,000 lbm/hr of oil available at 256°F.

Property values: c_{water} = 1.0 Btu/(lbm-°F)
c_{oil} = 0.54 Btu/(lbm-°F)

For this heat exchanger, the ratio of the LMTD counterflow to LMTD parallel flow (i.e., $LMTD_{counterflow}/LMTD_{parallel\ flow}$) is most nearly:

(A) 1.25
(B) 1.15
(C) 1.05
(D) 1.0

509. A 6.0-in.-O.D. tube with a wall thickness of 0.10 in. and an axial length of 20 ft is located in a roofed building. The temperature of the ambient air is 65°F, and the convective heat-transfer coefficient is 5.0 Btu/(hr-ft^2-°F).

The outer surface temperature of the pipe is 450°F. The emissivity for the pipe is 0.5. The total rate of heat loss (Btu/hr) would be most nearly:

(A) 93,000
(B) 77,000
(C) 62,000
(D) 16,000

510. To save on heating costs, a business is considering adding insulating shades to the 200 ft^2 of single-pane windows in its historic office building, as shown in the figure. Neglecting radiation, the reduction in heat loss (Btu/hr) that would be achieved by adding the shades is most nearly:

(A) 11,200
(B) 8,300
(C) 6,400
(D) 300

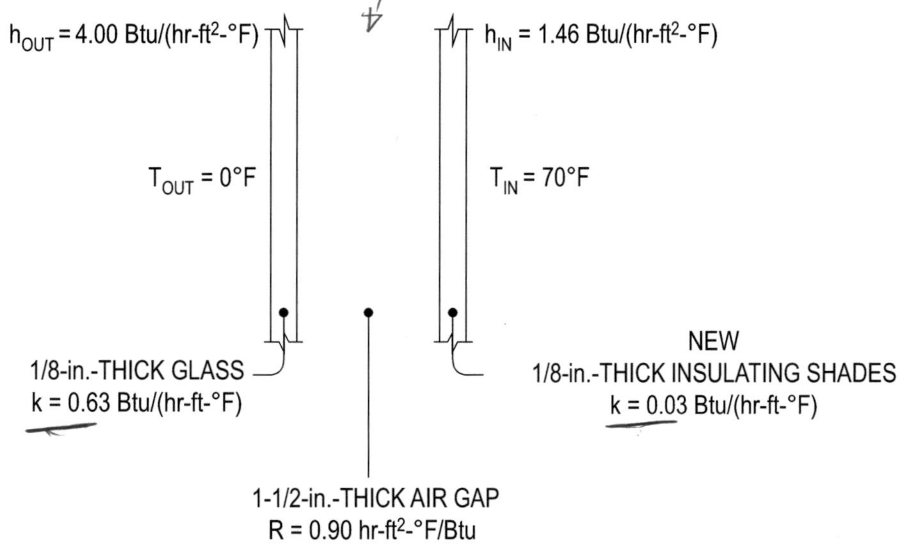

511. Below is a schematic diagram of the compressor of an aircraft gas turbine. A portion of the air that enters at State 1 is extracted at State 2, cooled and then expanded in a turbine to be used for cabin pressurization. The balance of the air leaves at State 3. Known properties (stagnation pressures and temperatures) and other data are given with the diagram. The power (hp) required to drive the compressor is most nearly:

(A) 370
(B) 15,900
(C) 16,500
(D) 22,100

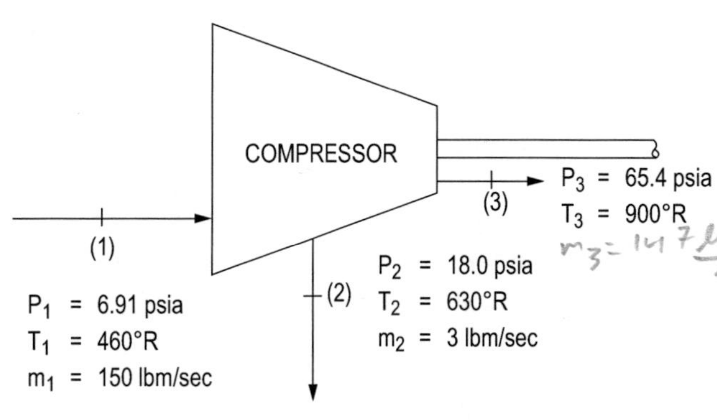

512. A system in which the resulting fluid (H_2O) is 7°F above saturation is shown in the figure below. The required flow rate (lbm/hr) at State 3 is most nearly:

(A) 3,500
(B) 3,850
(C) 4,225
(D) 5,163

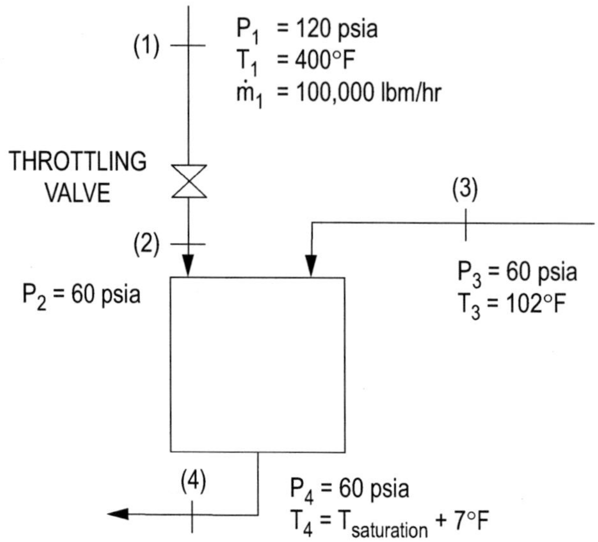

513. A nuclear reactor is housed in a leak-tight containment building. The building is designed to withstand the pressure and temperature resulting from an accidental breach of the pressure boundary of the reactor. During normal operating conditions, the reactor, which has a void volume of 9,000 ft^3, contains helium at 1,000 psia and 1,340°F. The free volume of the containment building is 225,000 ft^3, and it normally contains air at 14.65 psia at 140°F. Use 0°R as the reference temperature for calculating thermal properties.

	Helium	Air
c_p, Btu/(lbm-°F)	1.24	0.241
c_v, Btu/(lbm-°F)	0.745	0.172
k	1.667	1.400
M, lbm/(lb mol)	4	29

In an accident, helium from the reactor slowly leaks into the building. The operating crew of the plant is able to plug the leak before equilibrium pressure is reached. At this state, it is determined that the helium mass fraction in the building is 0.09091 and the building pressure is 30 psia (building air mass is 15,000 lbm). The mass (lbm) of helium released into the building is most nearly:

(A) 1,364
(B) 1,500
(C) 1,650
(D) 1,800

GO ON TO THE NEXT PAGE

514. A shell-and-tube-type closed feedwater heater is used to raise the temperature of water entering a boiler. The steam and feedwater conditions entering the feedwater heater are shown below. If steam leaves the feedwater heater as saturated liquid at 1,000 psia and the feedwater temperature rise across the heater is 80°F, the steam flow (lbm/hr) is most nearly:

(A) 11,000
(B) 35,000
(C) 134,000
(D) 248,000

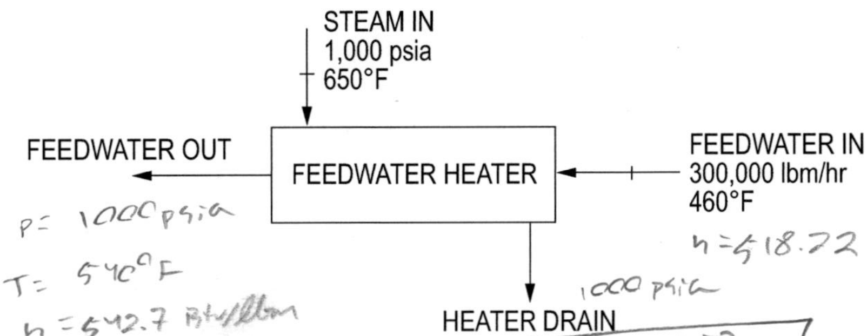

515. Fuel is burned in a furnace with 12% excess air to minimize the danger of partially oxidized product being formed. The average molecular weight of the fuel is 126. The overall complete **stoichiometric** combustion reaction is:

$$C_9H_{18} + 13.5\ O_2 + 50.8\ N_2 \rightarrow 9\ CO_2 + 9\ H_2O + 50.8\ N_2$$

The mass of air (lbm) required per pound of fuel for complete combustion with 12% excess air is most nearly:

(A) 3.8
(B) 14.8
(C) 16.5
(D) 18.3

516. The turbine efficiency from State 1 to State 2 is most nearly:

(A) 75%
(B) 79%
(C) 82%
(D) 87%

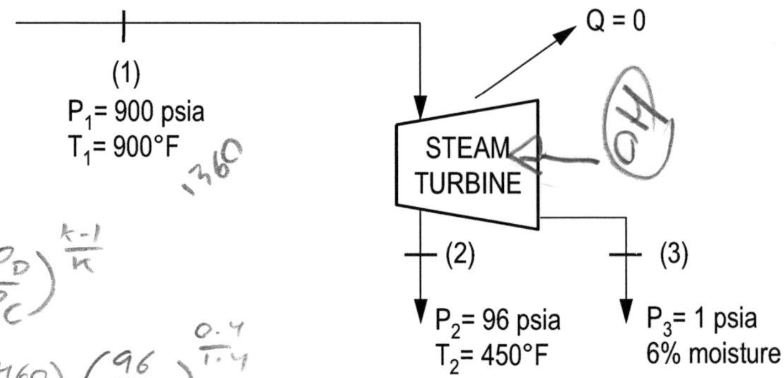

517. Assume an engine operates on an open ideal air-standard Brayton cycle. Air enters the compressor at 14.7 psia and 80°F. The pressure ratio is 6.5:1 and leaves the combustor at 1,550°F. The combustor heat addition (Btu/lbm) is most nearly:

(A) 150
(B) 210
(C) 260
(D) 330

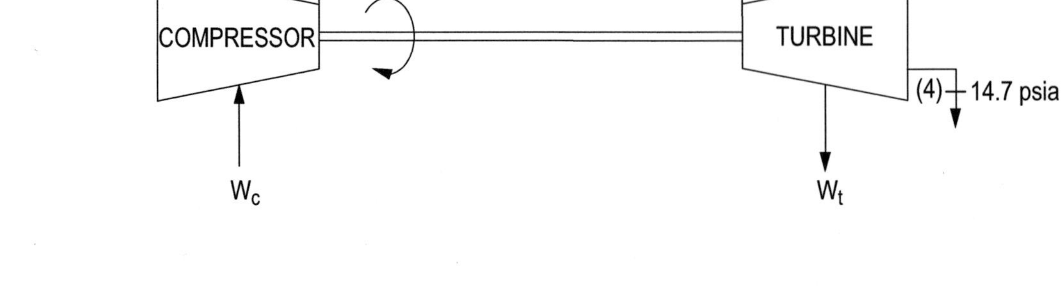

518. A psychrometric chart is provided on page 89 for your possible use in answering this question.

Air leaves an air-conditioning system at 58°F db and 90% rh with a flow rate of 10,000 cfm (State 1). It is mixed with 2,000 cfm of outside air at 95°F db and 47% rh (State 2). The entire process takes place at standard atmospheric pressure. The specific humidity or humidity ratio (lbm of water per lbm of dry air) of the air after the two streams are mixed is most nearly:

(A) 0.0103
(B) 0.0118
(C) 0.0131
(D) 0.0145

519. A centrifugal pump is sized to deliver 200 gpm at a head of 60 ft of water. It is operating at 1,750 rpm. The pump efficiency is 75%. When the speed of the pump has been reduced to 1,000 rpm, the head (ft of water) is most nearly:

(A) 8.5
(B) 11
(C) 20
(D) 34

520. A centrifugal pump operating with a pump head of 50 psi is used to transport 25 gpm of acetone at 78°F. The vapor pressure of acetone at this temperature is 5.6 psia, viscosity is 0.3 cP, and density is 49 pcf (sp. gr. is 0.79). The pump efficiency is 70%. The input shaft horsepower (hp) to drive the pump is most nearly:

(A) 0.12
(B) 0.73
(C) 1.04
(D) 1.35

521. Steam enters an adiabatic turbine at 1,000 psia and 900°F and leaves at 2 psia. The turbine efficiency is 80%. The steady-state work (Btu/lbm) developed by the turbine is most nearly:

(A) 325
(B) 390
(C) 409
(D) 490

522. A turbofan engine is operating on a thrust stand at sea level with an ambient temperature of 80°F and an ambient pressure of 14.7 psia. Known data for the fan-compressor portion are provided in the figure. The input power (hp) required to drive the combined fan-compressor is most nearly:

(A) 19,400
(B) 26,800
(C) 87,100
(D) 112,000

Compressor mass rate of flow	100 lbm/sec
Fan mass rate of flow	400 lbm/sec
Compressor pressure ratio	10
Fan pressure ratio	1.3
Fan and compressor efficiencies	85%

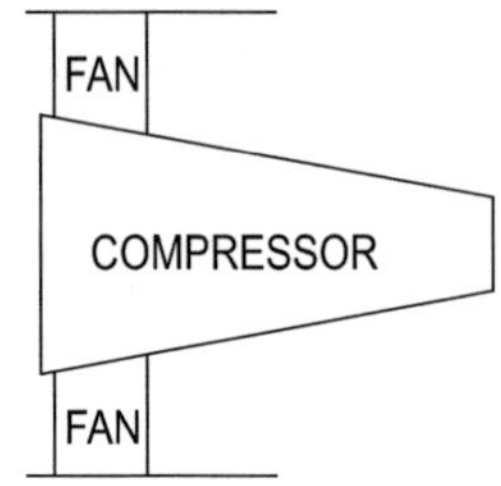

523. Known data for a steam generator are given below. The fuel is a mixture of coal and refuse-derived fuel as follows:

Fuel	HHV (Btu/lbm)	$\dot{m}$ (lbm/hr)
Coal	10,800	unknown
Refuse-derived	5,500	4,600

If the mass rate of flow of water through the steam generator is 90,000 lbm/hr and 75% of the fuel HHV is transferred to the water, the rate at which coal (lbm/hr) must be supplied is most nearly:

(A) 15,365
(B) 10,680
(C) 7,425
(D) 4,980

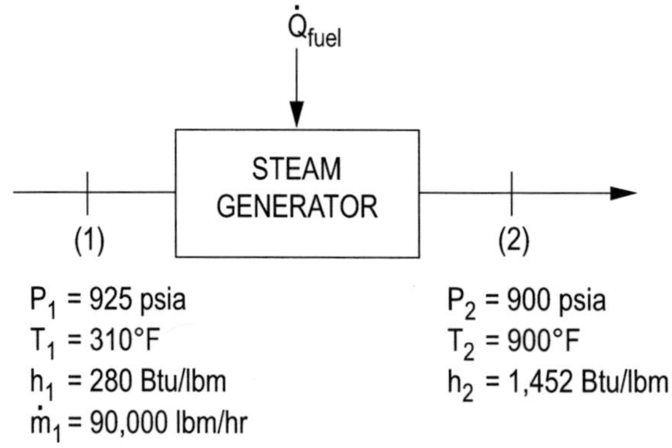

524. A 350-in^3-displacement V8 engine delivers 200 hp while running at 4,200 rpm. At these conditions the brake-specific fuel consumption is 0.65 lbm/(hp-hr), and the engine friction power is 50 hp. Fuel density is 42.8 lbm/ft^3. The fuel consumption (gpm) is most nearly:

(A) 0.26
(B) 0.38
(C) 0.48
(D) 3.00

525. Known data for a closed regenerative gas turbine cycle are given in the figure. Carbon dioxide [c_p = 0.203 Btu/(lbm-°R), c_p/c_v = 1.285, M = 44 lbm/lbmol] is the working fluid. Assuming constant specific heats throughout, the heat added (Btu/lbm) from an external source is most nearly:

(A) 108
(B) 137
(C) 157
(D) 255

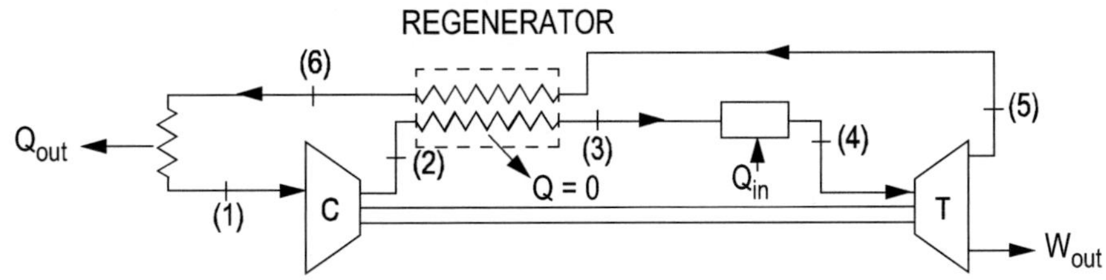

T_1 = 760°R
T_4 = 2,360°R

$P_1 = P_5 = P_6$ = 29.0 psia
$P_2 = P_3 = P_4$ = 116.0 psia

Compressor efficiency	80%
Turbine efficiency	85%
Regenerator effectiveness	80%

526. Given a hydronic piping system and a number of pumps, each having performance characteristics as shown in the figure below, the **minimum** number of available pumps needed to supply the required flow at system operation Point A is:

(A) two
(B) three
(C) four
(D) Cannot be done with the available pumps

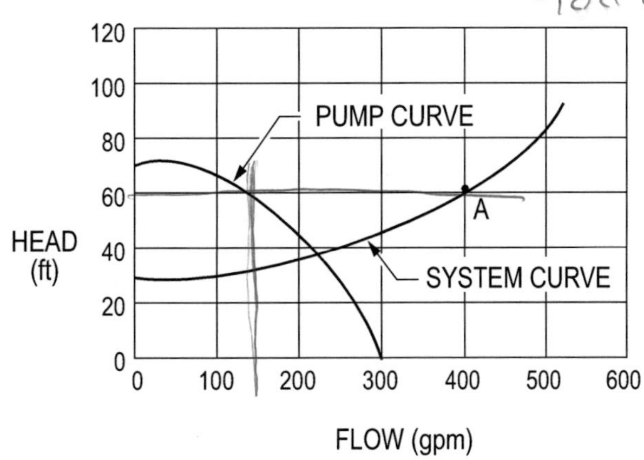

527. The exhaust gases from a gas turbine are used as the energy source for a two-pressure steam power plant as shown in the figure. The exhaust gases, with an average c_p of 0.252 Btu/(lbm-°R) and a molecular weight of 29.0 lbm/lb mol, enter the heat exchanger at a temperature of 1,036°F, a pressure of 14.9 psia, and a flow rate of 3,100 lbm/sec. The mass rate of flow of H_2O is 289 lbm/sec through the high-pressure turbine and 340 lbm/sec through the low-pressure turbine. Other known values are shown in the table below.

The volumetric flow rate (cfm) of the exhaust gases at State A is most nearly:

(A) 115×10^3
(B) $4{,}800 \times 10^3$
(C) $6{,}900 \times 10^3$
(D) $16{,}600 \times 10^3$

State	Pressure (psi)	Temperature (°F)	Other
1	1,050	960	—
2	94	420	—
3	94		—
4	0.50		Quality = 85%
5	0.50		Saturated Liquid
6	94		—
7	94		Quality = 15%
8	94		Saturated Vapor
9	94		Saturated Liquid
10	1,050		

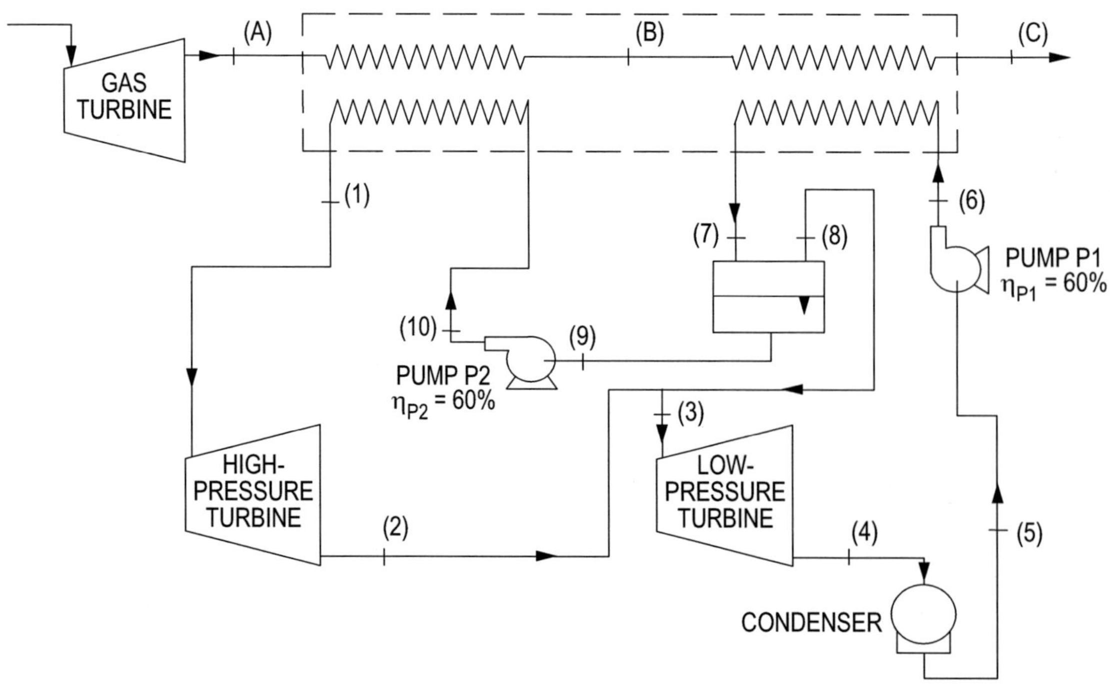

528. A piping system includes three pumps, each having the performance characteristics shown in the following graph. The supply head (ft) that three pumps operating in parallel will deliver to this hydronic system is most nearly:

(A) 210
(B) 65
(C) 38
(D) 30

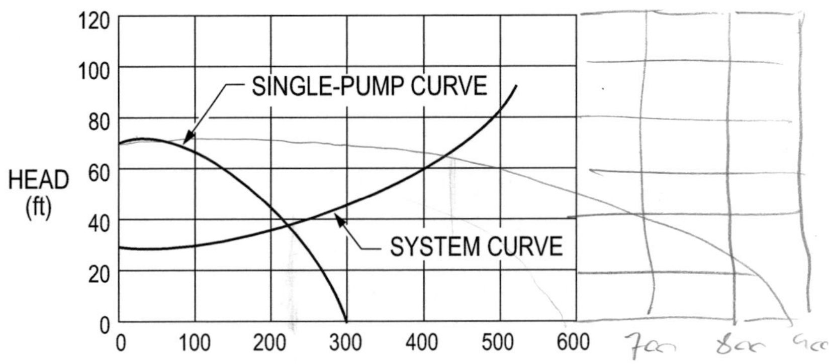

529. A motor-driven pump draws water from a large reservoir, as shown in the figure below. The water level in the reservoir remains constant. The pump is required to move 300 gpm. The suction pipe is 4-in. Schedule 40 black steel, and the discharge pipe is 3-in. Schedule 40 black steel. The large reservoir contains water at 193°F and 10 psia. The pressure at the pump discharge must be 120 psig.

Water properties at 193°F:

Density, ρ	60.3 lbm/ft^3
Kinematic viscosity, υ	0.356×10^{-5} ft^2/sec
Vapor pressure	23.06 ft of fluid

Pipe sizes:

3 in. nominal is 3.068 in. I.D.
4 in. nominal is 4.026 in. I.D.

Assuming the friction loss to be 2.84 ft, the net positive suction head, or NPSH, (ft) is most nearly:

(A) 9.5
(B) 10.5
(C) 12.5
(D) 14.5

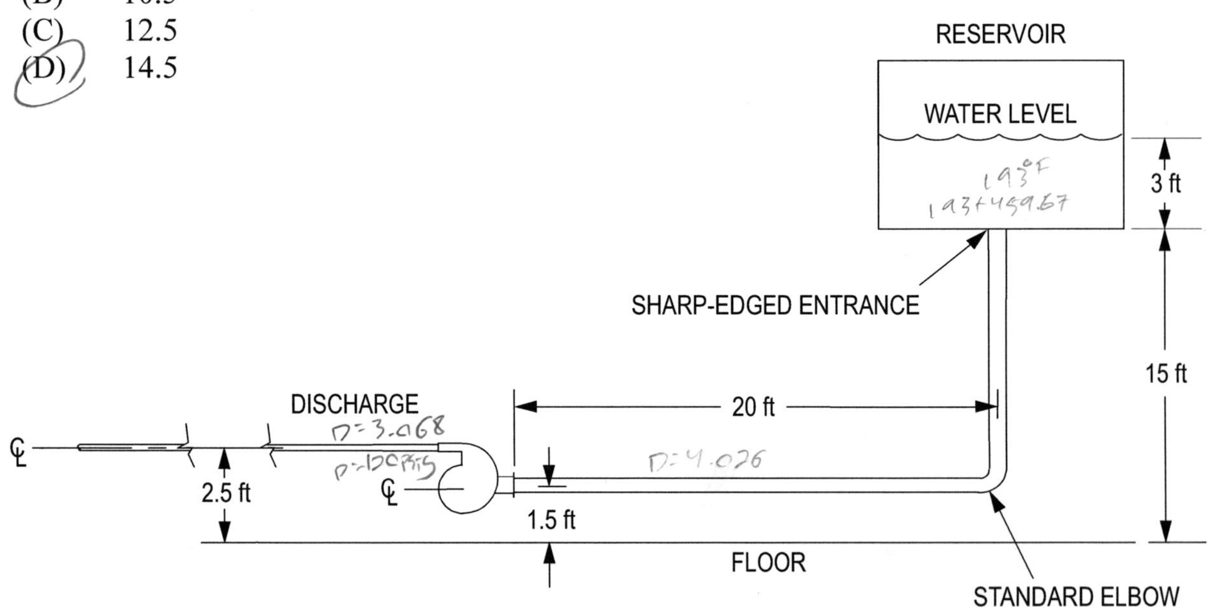

NOT TO SCALE

$NPSH = h_{atm} + h_{z(s)} - h_{f(s)} - h_{vp}$

$= \frac{P_a}{\rho g} + h_{z(s)} - h_{f(s)} - h_{vp}$

$= \left[\frac{\frac{10\ lbf}{in^2} \cdot \frac{144\ in^2}{ft^2} \cdot \frac{32.2\ lbm\text{-}ft}{lbf \cdot s^2}}{\frac{60.3\ lbm}{ft^3} \cdot \frac{32.2\ ft}{s^2}} \right] + (18 - 1.5)\ ft - 2.84 - 23.06$

$NPSH = 14.5$

GO ON TO THE NEXT PAGE

530. A motor-driven pump draws water from a large reservoir, as shown in the figure below. The water level in the reservoir remains constant. The pump is required to move 300 gpm. The suction pipe is 4-in. Schedule 40 black steel, and the discharge pipe is 3-in. Schedule 40 black steel. The large reservoir contains water at 193°F and 10 psia. The pressure at the pump discharge must be 120 psig.

Water properties at 193°F:

Density, ρ	60.3 lbm/ft^3
Kinematic viscosity, υ	0.356×10^{-5} ft^2/sec
Vapor pressure	23.06 ft of fluid

Pipe sizes:

3 in. nominal is 3.068 in. I.D.
4 in. nominal is 4.026 in. I.D.

Assume the following:

$V_{4\ in}$	7.6 fps
$V_{3\ in}$	13.0 fps
Darcy friction factor	0.018
K_{elbow}	0.9
$K_{entrance}$	0.5

For the conditions given, the total differential head (ft) of liquid developed by the pump is most nearly:

(A) 288
(B) 298
(C) 303
(D) 312

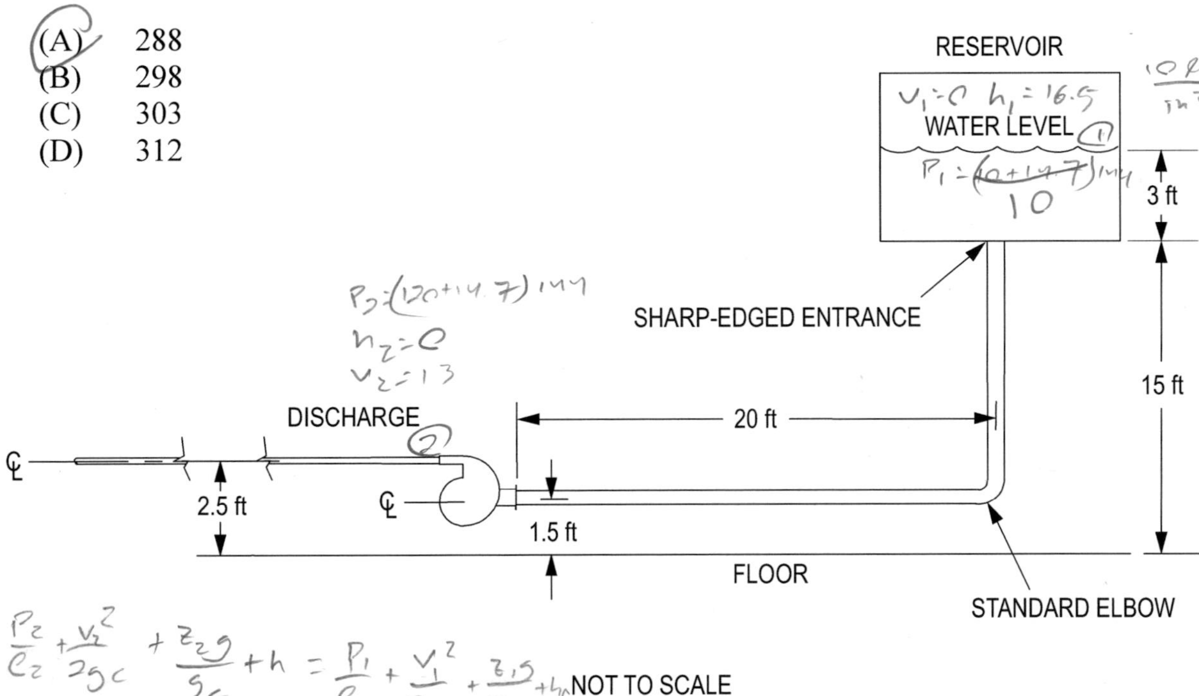

GO ON TO THE NEXT PAGE

531. The outside air conditions are 40°F and 60% rh. The space design temperature is 75°F and 50% rh. The outside airflow rate is 3,000 cfm. The constant volume system has a supply air design flow of 11,500 cfm. The mixed-air temperature under these conditions is most nearly:

(A) 50°F
(B) 56°F
(C) 60°F
(D) 66°F

532. A motorcycle is equipped with a disc brake system consisting of a single-circuit master cylinder and caliper as shown. The caliper piston has a diameter of 30 mm. The master cylinder has a diameter of 20 mm. A reaction force, R, of 200 N is required at the caliper. The input force F (N) required at the master cylinder is most nearly:

(A) 90
(B) 130
(C) 200
(D) 450

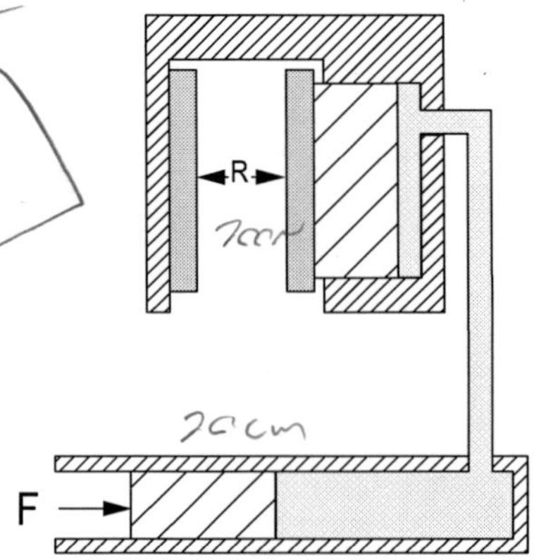

533. A cogeneration facility generates electricity and supplies steam to a chemical process plant, as shown in the figure below.

The following data are known:

Main steam supply	2,500 psia, 1,000°F
FW, Feedwater supply to boiler	3,000 psia, 400°F
Extraction steam supply	300 psia, 500°F

Extracted steam to closed feedwater heater leaves the heater as saturated liquid at 300 psia.

Extraction steam is used to heat feedwater in a closed feedwater heater. Condensate leaves the heater as saturated liquid. The quantity of steam (lbm of steam per lbm of feedwater) needed to heat the feedwater from 110°F to 400°F is most nearly:

(A) 0.10
(B) 0.22
(C) 0.34
(D) 0.62

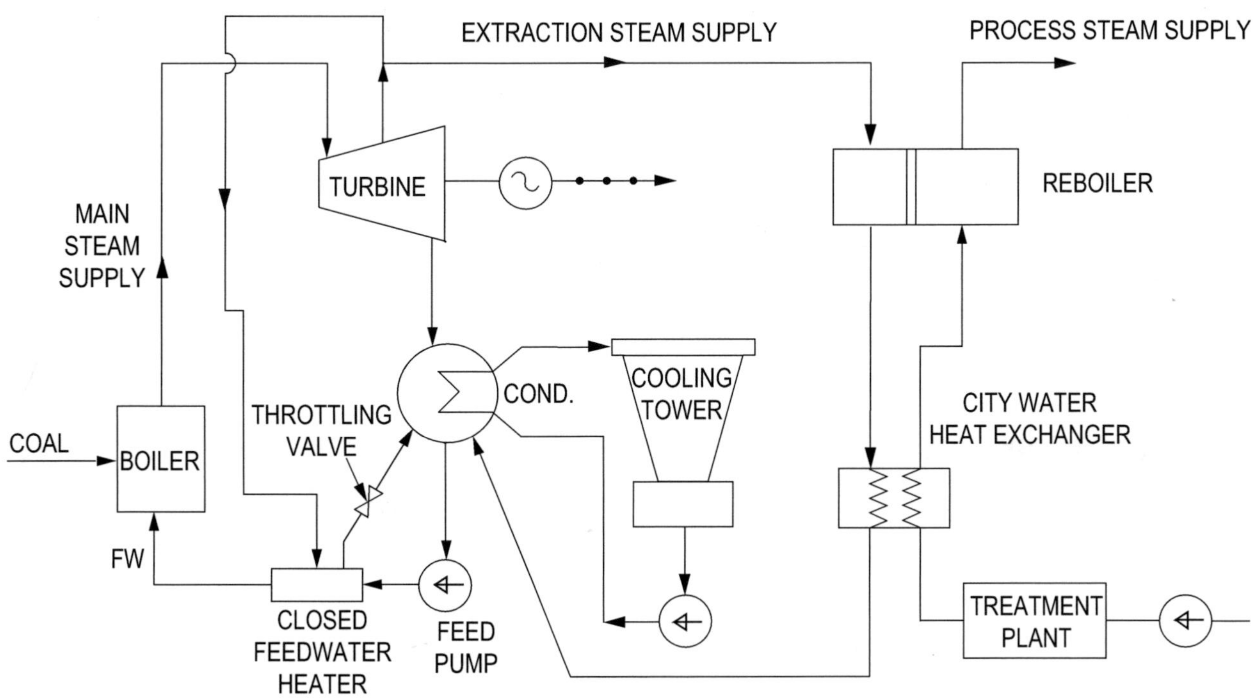

534. A pitot tube measurement at a point that represents the average velocity indicates a head of 3 in. of water for air (ρ = 0.075 lbm/ft^3) flowing in a circular duct with a diameter of 12 in. The volumetric flow rate (cfm) of air is most nearly:

(A) 1,725
(B) 3,860
(C) 5,450
(D) 18,900

535. For a given set of fixed steam conditions ($\dot{m}$, P, T) at the inlet to the high-pressure turbine of a fossil fuel-fired steam power plant with reheat between sections of a two-stage turbine, which of the following is a result of adding reheat?

(A) Increased moisture content at the exit from the low-pressure turbine
(B) Decreased net power output
(C) Decreased amount of energy exchanged in the condenser
(D) Increased CO_2 emissions

536. The psychrometric processes for air supplied to an auditorium are represented as Lines A and B on the accompanying psychrometric chart. Line A is the room load line, and Line B is the coil load line. The coil apparatus dew-point temperature is most nearly:

(A) 50°F
(B) 53°F
(C) 58°F
(D) 63°F

536. (Continued)

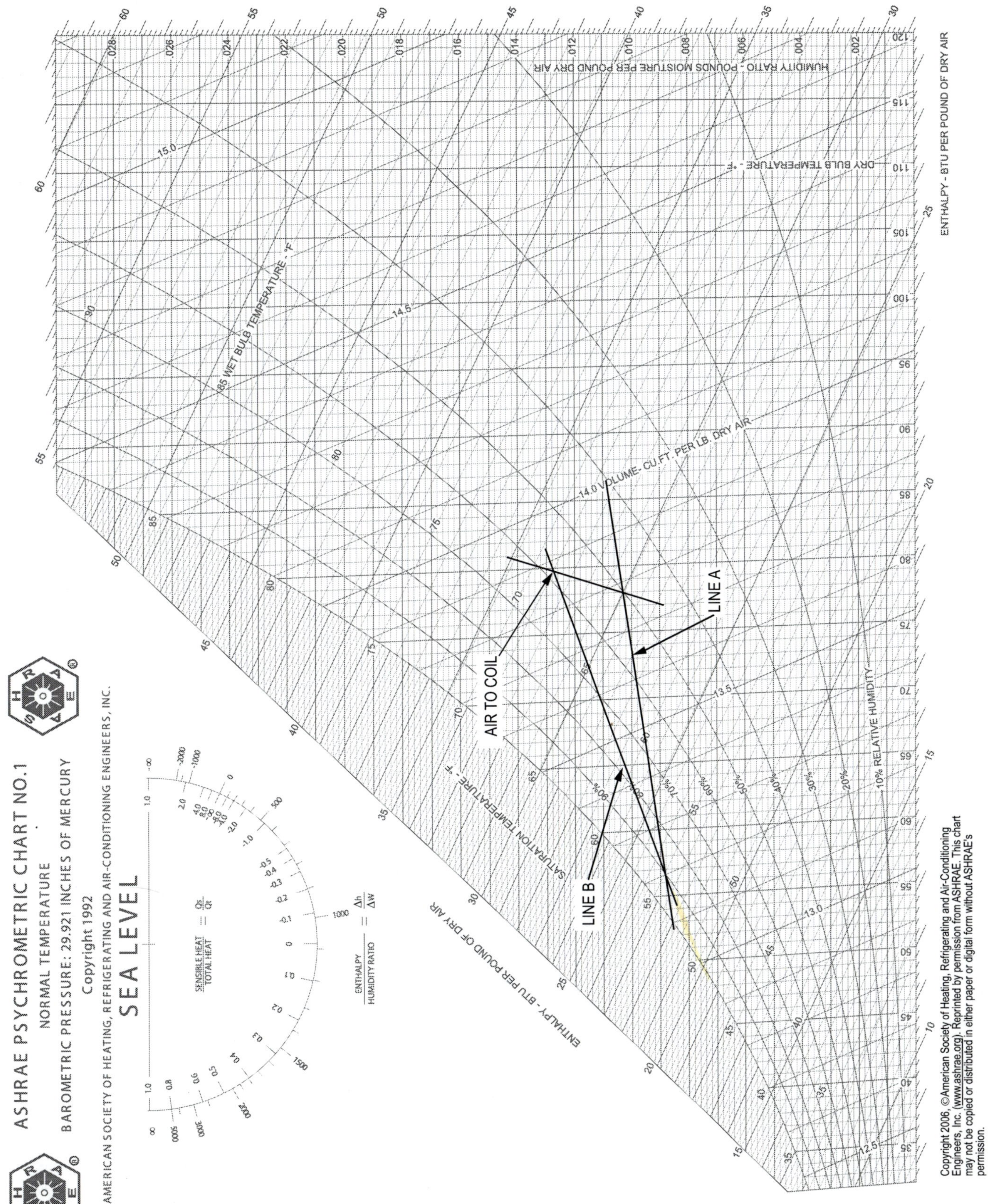

537. A pellet fuel consisting of waste paper and plastic is used to displace a high-sulfur waste coal (gob coal) on a constant Btu basis in a 50-MW circulating fluidized bed (CFB) combustor. The two fuels are characterized in the following table:

Parameter	Gob Coal	Pellets	Ash
Heating Value (Btu/lbm)	5,500	6,550	
Sulfur (%, wt)	3.00	0.15	
Ash (%, wt)	40.0	10.0	
Disposal cost ($/ton*)	–	–	$10

*1 ton = 2,000 lbm

The reduction in ash disposal costs (per ton of coal displaced) from using pellets is most nearly:

(A) $0.84
(B) $2.81
(C) $3.00
(D) $3.16

~~50 MW~~

2000 lbm · 5500 Btu/lbm = 11 000 000 Btu

2000 lbm · 0.40 = 800

800/2000 · $10 = $4

11000000 Btu · lbm/6550 Btu = 1679 lbm

1679 lbm · 0.10 = 167.9389

167.9389/2000 · $10 = 0.83

  GO ON TO THE NEXT PAGE

538. The following information is known about a new power plant.

Size	400 MW
Heat rate	11,010 Btu/kWh
Fuel	coal
Higher heating value of coal	14,100 Btu/lbm

Ultimate "as-fired" analysis (% weight) of coal	
Carbon	80.5%
Hydrogen	4.4%
Sulfur	1.9%
Oxygen	2.4%
Nitrogen	1.3%
Water	3.2%
Ash	6.3%
Total	100.0%

The mass (lbm CO_2/day) produced when the plant operates at full load for 24 hours and all carbon goes to carbon dioxide is most nearly:

(A) 1.646×10^6
(B) 6.034×10^6
(C) 22.13×10^6
(D) 39.29×10^6

539. The pressure vessel of a gas-fired low-pressure steam boiler is required to meet codes and standards published by:

(A) AGA
(B) UL
(C) ASME
(D) NFPA

540. A thin-walled pressure vessel is to be made from a quenched and tempered forging conforming to ASTM 372 (relevant portions reproduced below). To comply with the standard, the finished product must be examined using:

(A) liquid penetrant examination
(B) digital radiography
(C) ultrasonics
(D) magnetic particle testing

Designation: A 372/A 372M – 99 **An American National Standard**

AMERICAN SOCIETY FOR TESTING AND MATERIALS
100 Bar Harbor Dr., West Conshohocken, PA 19428
Reprinted from the Annual Book of ASTM Standards. Copyright ASTM

Standard Specification for Carbon and Alloy Steel Forgings for Thin-Walled Pressure Vessels[1]

2. Referenced Documents

2.1 ASTM Standards:

A 275/A 275M Test Method for Magnetic Particle Examination of Steel Forgings[3]
A 370 Test Methods and Definitions for Mechanical Testing of Steel Products[4]
A 388/A 388M Practice for Ultrasonic Examination of Heavy Steel Forgings[3]
A 530/A 530M Specification for General Requirements for Specialized Carbon and Alloy Steel Pipe[5]
A 788 Specification for Steel Forgings, General Requirements[3]
E 112 Test Methods for Determining the Average Grain Size[6]
E 165 Practice for Liquid Penetrant Examination[7]
E 290 Test Method for Semi-Guided Bend Test for Ductility of Metallic Materials[6]
E 433 Reference Photographs for liquid Penetrant Inspection[7]

7. Magnetic Particle Examination

7.1 All quenched and tempered forgings shall be subject to magnetic particle examination after heat treatment. The methods used shall be in accordance with Test Method A 275/A 275M.

7.2 Only direct current or rectified alternating (full or half wave) current shall be used as the electric power source for any of the magnetizing methods.

7.3 The forgings shall be free from linear indications indicative of cracks. In case of doubt as to the type of condition giving rise to linear magnetic particle indications, they may be inspected for Type II indications referenced in Reference Photographs E 433 when inspected by the liquid penetrant method in accordance with Practice E 165. Alternatively, some other suitable method for classifying the indications may be used as agreed upon between the manufacturer and the purchaser.

---

[1]This specification is under the jurisdiction of ASTM Committee A-1 on Steel, Stainless Steel and Related Alloys and is the direct responsibility of Subcommittee A01.06 on Steel Forgings and Billets.

Current edition approved March 10, 1999. Published June 1999. Originally published as A 372-53T. Last previous edition A 372/A 372M-95.

[3]Annual Book of ASTM Standards, Vol. 01.05.
[4]Annual Book of ASTM Standards, Vol. 01.03.
[5]Annual Book of ASTM Standards, Vol. 01.01.
[6]Annual Book of ASTM Standards, Vol. 03.01.
[7]Annual Book of ASTM Standards, Vol. 03.03.

From *American Society for Testing and Materials, Annual Book of ASTM Standards*

This completes the afternoon session. Solutions begin on page 67.

MECHANICAL
MORNING SOLUTIONS

Answers to the Mechanical MORNING Sample Questions

Detailed solutions for each question begin on the next page.

101	C	121	A
102	A	122	B
103	D	123	C
104	A	124	D
105	B	125	D
106	C	126	B
107	D	127	B
108	A	128	A
109	C	129	C
110	D	130	D
111	C	131	B
112	D	132	C
113	D	133	C
114	D	134	B
115	D	135	B
116	C	136	B
117	C	137	C
118	C	138	D
119	B	139	C
120	D	140	D

MECHANICAL MORNING SOLUTIONS

101. Poisson's ratio is a measure of how a material deforms when acted upon by a force. It is applied in situations where the mechanical behavior of a machinery component is important.

THE CORRECT ANSWER IS: (C)

102. The coefficient of performance (COP) is a measure of energy efficiency of refrigeration systems.

THE CORRECT ANSWER IS: (A)

103. Annualized purchase price over 10 years $90,000 (A/P, 10%, 10) = 0.1627
Annualized salvage value after 10 years $10,000 (A/F, 10%, 10) = 0.0627
Annual operating cost = (300 days/yr)(16 hr/day)($2.50/hr) = $12,000
Total annual cost = $90,000(0.1627) – $10,000(0.0627) + $12,000 = $26,016

THE CORRECT ANSWER IS: (D)

104. 80 hours – 10% lost production time = 72 hours
72 hours × 60 units/hour = 4,320 units ⇒ 1 scrap/120 units × 4,320 units = 36 scrap
4,320 – 36 units = 4,284 units

THE CORRECT ANSWER IS: (A)

105. If Task C is delayed by 3 days, but has 2 days to spare while waiting on Task D, then the total compensation delay will be 1 day of delivery time.

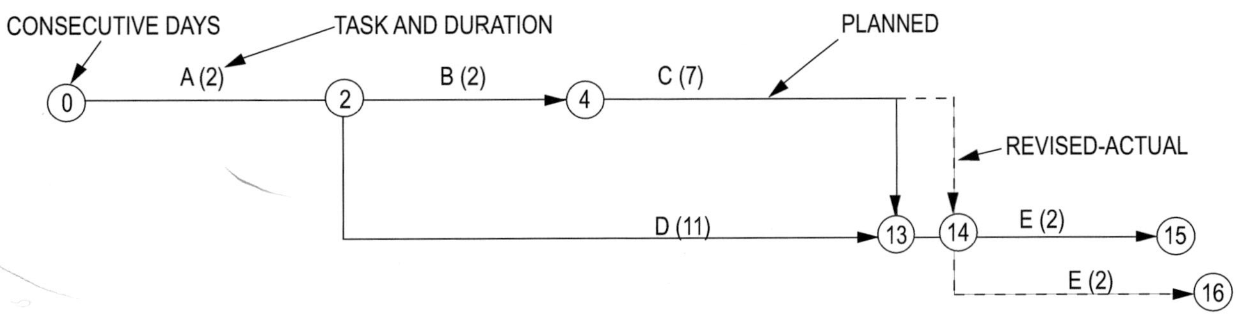

THE CORRECT ANSWER IS: (B)

en the maximum cost of \$36 per drum, the maximum amount of active ingredient is obtained by using as much as allowed of the cheaper formulation, and then topping off with the more costly formulation. At \$10/2 = \$5 per 100 units of active ingredient, Formulation 2 is less expensive. The maximum amount of Formulation 2 is 2 lb, which costs \$20 and gives 400 units of active ingredient. This leaves \$36 – \$20 = \$16 to spend on Formulation 1. This will buy 2 lb, which contains 2 lb × 100 units/lb = 200 units of active ingredient. This gives a total of 400 + 200 = 600 units, which is the maximum achievable within the constraints.

THE CORRECT ANSWER IS: (C)

107.

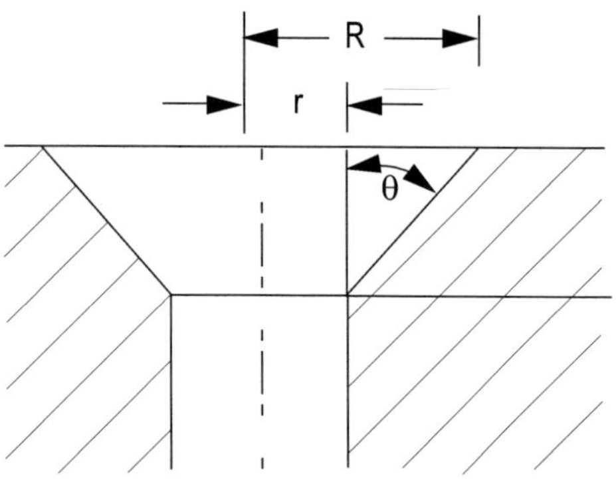

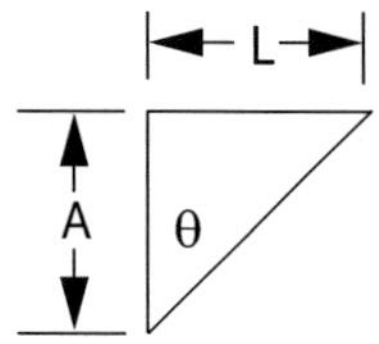

$L = R - r = 0.39 - 0.156 = 0.234$ in.

$$A = \frac{L}{\tan\theta} = \frac{0.234}{\tan 41^\circ} = 0.269 \text{ in.}$$

Detail of outer part of countersink

$R = \varnothing/2 = 0.78 \text{ in.}/2 = 0.39$ in.
$r = \text{I.D.}/2 = 5/16 \text{ in.}/2 = 0.156$ in.
$\theta = 82^\circ/2 = 41^\circ$

THE CORRECT ANSWER IS: (D)

108.

1. Determine minimum centerline of holes	$℄ = 2.4 - 1.2 - 2(0.05) = 1.1$
2. Determine the maximum outside diameter of holes	$D_{max} = 0.9$
3. Determine minimum distance between surfaces	$1.1 - 0.9 = 0.20$

THE CORRECT ANSWER IS: (A)

109. According to Marks' *Standard Handbook for Mechanical Engineers,* 9th edition, p. 15-7, "the combined resistance of a number of series-connected resistors is the sum of their separate resistances."

$$R_e = R_1 + R_2 + R_3$$

THE CORRECT ANSWER IS: (C)

110. $I = \dfrac{(sf)(P)}{(V)\eta(pf)}$

$I = \dfrac{(1.0)(7{,}457)}{(230)(0.65)(0.80)} = 62.3\text{ A}$

Service factor, sf = 1.0 at full load
Power, P = 10 hp × 0.7457 kW/hp × 1,000 W/kW = 7,457 W
Voltage, V = 230 V
Efficiency, η = 0.65
Power factor, pf = 0.80

THE CORRECT ANSWER IS: (D)

111. Gravitational potential energy for a mass at a height is $E = \dfrac{mgh}{g_c}$.

The rate of change in that energy $\dot{E} = \dfrac{\dot{m}gh}{g_c}$

$\dot{m} = \rho\dot{Q} = (62.4\text{ lbm/ft}^3)(4{,}000{,}000\text{ ft}^3/\text{min})(1\text{ min}/60\text{ sec}) = 4{,}200{,}000\text{ lbm/sec}$

$\dot{E} = \dfrac{(4{,}200{,}000\text{ lbm/sec})(32.2\text{ ft/sec}^2)(170\text{ ft})}{32.2\text{ lbm-ft/(lbf-sec}^2)} = 707{,}000{,}000\text{ ft-lbf/sec}$

$= 707{,}000{,}000\text{ ft-lbf/sec} \times \dfrac{1.355\text{ W}}{1\text{ ft-lbf/sec}} \times \dfrac{1\text{ MW}}{10^6\text{ W}} = 958\text{ MW}$

THE CORRECT ANSWER IS: (C)

112. 1 ksi = 6.8948 MPa

$1\sqrt{\text{in.}} = 0.15937\sqrt{\text{m}}$

$1\text{ ksi}\sqrt{\text{in.}} = 1.0989\text{ MPa}\sqrt{\text{m}}$

$35\text{ ksi}\sqrt{\text{in.}} = 38.46\text{ MPa}\sqrt{\text{m}}$

THE CORRECT ANSWER IS: (D)

113. Determine the elongation of a 6-in.-long, 1/2-in.-diameter rod when loaded with a 2,300-lbf force. Elongation of this rod is a linear function of stress from 0 to 70,0000 psi and equals 0.012 in./in. when stress equals 70,000 psi.

Stress = force/unit area = (2,300 lbf)/[(3.14159)(0.25 in. × 0.25 in.)]
= 11,700 psi for the rod as loaded.

Unit elongation = (11,700 psi/70,000 psi)(0.012 in./in. at S = 70,000 psi)
= 0.002 in./in.

For the 6-in.-long rod, the elongation is (6 in.)(0.002 in./in.) = 0.012 in.

THE CORRECT ANSWER IS: (D)

114. The von Mises failure criterion predicts failure where $\sigma' = S_{yt}$, where $\sigma' = \sqrt{\sigma_1{}^2 + \sigma_2{}^2 - \sigma_1\sigma_2}$ for biaxial loading, which is the case with this thin-walled cylinder. σ_1 is the hoop stress, and σ_2 is the longitudinal stress.

$$\sigma_1 = \frac{pr}{t} = \frac{3\text{ MPa} \times \dfrac{1{,}000{,}000\text{ Pa}}{1\text{ MPa}} \times \dfrac{1\text{ m}}{2}}{1\text{cm}\left(\dfrac{1\text{ m}}{100\text{ cm}}\right)} = 150\text{ MPa}$$

$$\sigma_2 = \frac{pr}{2t} = \frac{\sigma_h}{2} = 75\text{ MPa}$$

$$\sigma' = \sqrt{(150\text{ MPa})^2 + (75\text{ MPa})^2 - (150\text{ MPa})(75\text{ MPa})} = 130\text{ MPa}$$

$$FS = \frac{S_{yt}}{\sigma'} = \frac{450\text{ MPa}}{130\text{ MPa}} = 3.46$$

THE CORRECT ANSWER IS: (D)

115. Determine the minimum force, P, required to raise the 50-lbf weight.

For this situation, the ratio:

$$P/W = e^{f\theta}$$

where: f = the coefficient of friction, and
θ = the angle of wrap

Therefore:
$$P = We^{0.35 \times 3.14159}$$
$$= (50\text{ lbf})(e^{0.35 \times 3.14159})$$
$$= 150\text{ lbf}$$

THE CORRECT ANSWER IS: (D)

116. Angular frequency, $\omega = \sqrt{\dfrac{Kg_c}{m}} = \sqrt{\dfrac{(20\text{ lbf/in.})(12\text{ in./1 ft})[32.2\text{ lbm-ft/(lbf-sec}^2)]}{10\text{ lbm}}} = 27.8\text{ rad/sec}$

Linear frequency, $f = \dfrac{\omega}{2\pi} = \dfrac{27.8\text{ rad/sec}}{2\pi\text{ rad/cycle}} = 4.4\text{ cycles/sec}$

THE CORRECT ANSWER IS: (C)

117. $K = \dfrac{d^4 G}{8 D^3 N}$

where: d = wire diameter
D = spring diameter
N = number of active coils
G = shear modulus
K = spring constant

$N_{st} = N_{br}; \quad D_{st}^3 = D_{br}^3; \quad K_{st} = K_{br}$

$$\therefore \frac{d_{st}^4 G_{st}}{8 D_{st}^3 N_{st}} = \frac{d_{br}^4 G_{br}}{8 D_{br}^3 N_{br}} \Rightarrow d_{st}^4 G_{st} = d_{br}^4 G_{br} \Rightarrow d_{br} = d_{st}\sqrt[4]{G_{st}/G_{br}}$$

$$d_{br} = 0.15\sqrt[4]{\frac{11\times10^6}{5.5\times10^6}} = 0.18 \text{ in.}$$

THE CORRECT ANSWER IS: (C)

118. Using an iron-carbon diagram, Option (C) will cool to ferrite and pearlite.

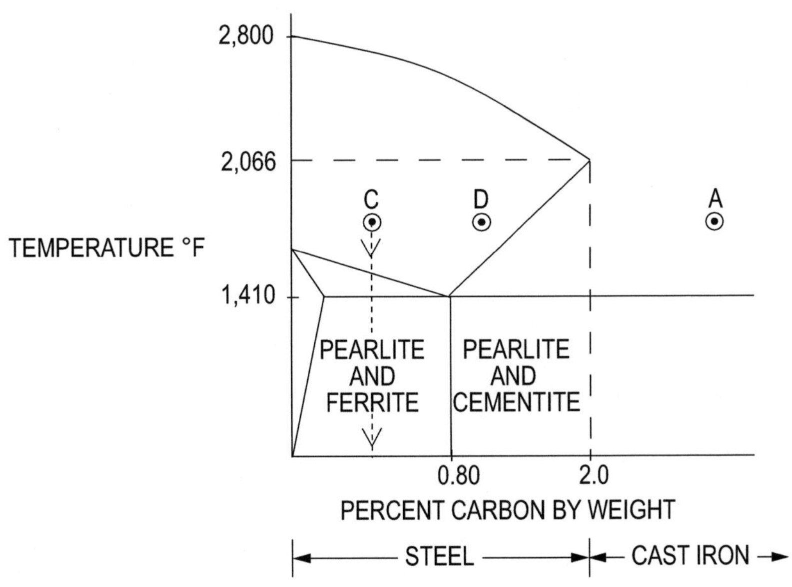

Option (B) is incorrect since AISI 304, a stainless steel, contains numerous alloying components in addition to the iron and carbon shown on this diagram. The last two digits of the AISI material code indicate the weight percent carbon. "95" would be to the right of 0.80 and is outside the pearlite/ferrite region. "40" is to the left of 0.80 and is at the middle of the pearlite/ferrite region, or about 50% of each. Therefore, AISI 1040 meets the requirements of the question.

THE CORRECT ANSWER IS: (C)

119. Fatigue life is expressed in number of cycles to failure and does not depend on the duration of each cycle (the cycle period). Since primary and alternating stresses and surface roughness do affect fatigue life, cycle period is least likely to affect fatigue life.

THE CORRECT ANSWER IS: (B)

120. Draw a free-body diagram.

Forces due to load:

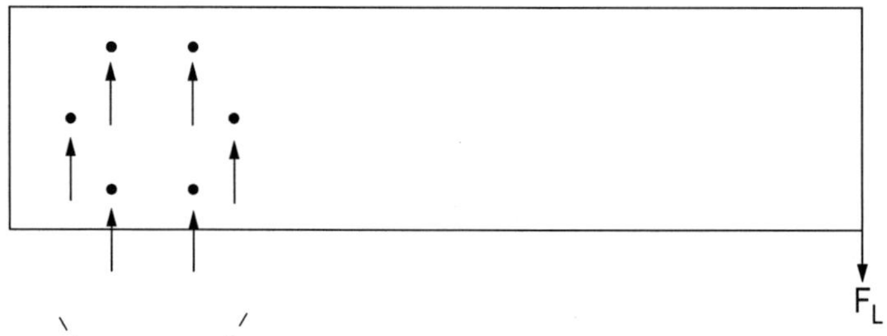

plus
Forces due to torsion:

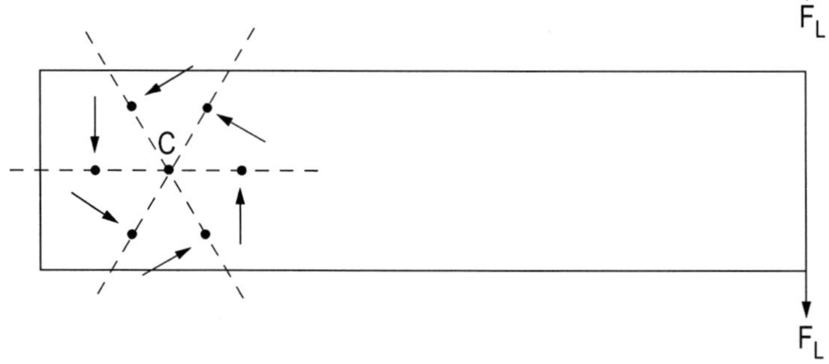

Add vectors:

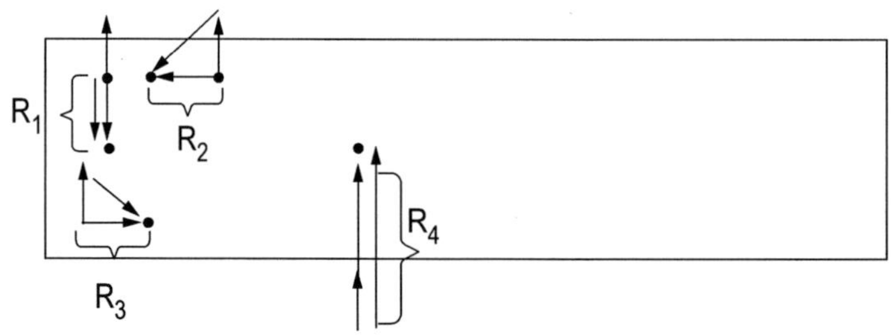

Bolt 4 has additive vectors and the greatest shear stress.

THE CORRECT ANSWER IS: (D)

121. This problem can be solved by referring to a normal shock table, on which all normal shock factors depend only on the upstream Mach number (for a given k value). With an upstream Mach number of 4.5, the normal shock table gives the downstream Mach number of 0.4236.

THE CORRECT ANSWER IS: (A)

122. Flow in = flow out

$$A_1V_1 + A_2V_2 = A_3V_3$$

$$A_1V_1 = \left(\pi \times \frac{10^2}{4}\right) \times 12 \text{ in./sec}$$

$$= 942 \text{ in}^3\text{/sec}$$

$$A_2V_2 = \left(\pi \times \frac{5^2}{4}\right) \times 120 \text{ in./sec}$$

$$= 2,356 \text{ in}^3\text{/sec}$$

$$A_3V_3 = 942 + 2,356$$

$$= 3,298 \text{ in}^3\text{/sec}$$

$$V_3 = 3,298 / \left(\pi \times \frac{12^2}{4}\right)$$

$$= 29.16 \text{ in./sec}$$

$$= 2.43 \text{ fps}$$

THE CORRECT ANSWER IS (B)

123. $Re = \dfrac{vD}{\upsilon}$

$$Re = \frac{(5 \text{ fps})(0.33 \text{ ft})}{1.615 \times 10^{-5} \text{ft}^2/\text{sec}} = 1.032 \times 10^5$$

THE CORRECT ANSWER IS: (C)

124. Using the pump similarity laws, power is directly proportional to the cube of the speed.

$$P_2 = P_1(S_2/S_1)^3$$

$$P_2 = 1,000 \text{ hp } (1,000 \text{ rpm}/1,750 \text{ rpm})^3$$

$$= 186 \text{ hp}$$

$$= 139 \text{ kW}$$

THE CORRECT ANSWER IS: (D)

125. $\dot{w} = \dot{m}\,\Delta H$, where $\Delta H = 60$ ft

$\dot{m} = \rho\,\dot{Q}$, where $\rho = 62.4\ \text{lbm/ft}^3$

$$\dot{Q} = \left(\frac{200\ \text{gal}}{\text{min}}\right)\left(\frac{1\ \text{ft}^3}{7.48\ \text{gal}}\right) = 26.7\ \text{ft}^3/\text{min}$$

$$\therefore \dot{m} = \left(\frac{62.4\ \text{lbm}}{\text{ft}^3}\right)\left(\frac{26.7\ \text{ft}^3}{\text{min}}\right) = 1{,}668\ \text{lbm/min}$$

$$\therefore \dot{w} = \left(\frac{1{,}668\ \text{lbm}}{\text{min}}\right)(60\ \text{ft}) = 100{,}080\ \text{ft-lbm/min}$$

$$\dot{w} = \left(\frac{100{,}080\ \text{ft-lbm}}{\text{min}}\right)\left(\frac{\text{min-hp}}{33{,}000\ \text{ft-lbm}}\right) = 3.03\ \text{hp}$$

$$\dot{w} = \frac{3\ \text{hp}}{(0.75)(0.83)} = 4.87\ \text{hp} = 3.63\ \text{kW}$$

$$\$ = (3.63\ \text{kW})(8{,}000\ \text{hr})\left(\frac{\$0.06}{\text{kWh}}\right) = \$1{,}742$$

THE CORRECT ANSWER IS: (D)

126. $\Delta P = f\dfrac{L}{D}\rho\dfrac{V^2}{2g_c}$

D = 1.063 in.
V = 5 ft/sec
$f = 0.025$
L = 100 ft = 1,200 in.
$g_c = 32.2\ \text{ft/sec}^2$

$$\Delta P = (0.025)\left(\frac{1{,}200\ \text{in.}}{1.063\ \text{in.}}\right)\left(\frac{62.4\ \text{lbm}}{\text{ft}^3}\right)\left(\frac{25\ \text{ft}^2}{\text{sec}^2}\right)\left(\frac{\text{sec}^2}{64.4\ \text{ft}}\right)$$

$$\Delta P = \left(\frac{684\ \text{lb}}{\text{ft}^2}\right)\left(\frac{1\ \text{ft}^2}{144\ \text{in}^2}\right) = 4.75\ \text{lb/in}^2 = 4.75\ \text{psi}$$

THE CORRECT ANSWER IS: (B)

127. Darcy Weisbach equation: $h_f = \dfrac{f\,Lv^2}{2\,Dg}$

$L_{eq,\,B} = 100\text{ ft} + (4 \times 5.2\text{ ft}) + 29\text{ ft} = 150\text{ eq. ft}$

In this configuration, the pressure drop across Branch A is equal to the pressure drop across Branch B,

so $h_{f,\,A} = h_{f,\,B} \Rightarrow \dfrac{f\,L_A V_A^2}{2\,Dg} = \dfrac{f\,L_B V_B^2}{2\,Dg} \Rightarrow V_A = V_B\sqrt{\dfrac{L_B}{L_A}} = V_B\sqrt{\dfrac{150\text{ ft}}{80\text{ ft}}} = 1.37\,V_B$

Flow is proportional to V, thus $q_A = 1.37\,q_B$

$20\text{ gpm} = q_A + q_B$

$20\text{ gpm} = 1.37\,q_B + q_B$

$q_B = 8.4\text{ gpm}$

THE CORRECT ANSWER IS: (B)

128. $\eta_{th,\,Carnot} = \dfrac{T_{high} - T_{low}}{T_{high}}$ T in absolute temperature scale.

$$\eta_{th,\,Carnot} = \frac{(400+460)-(250+460)}{(400+460)} = \frac{860^\circ R - 710^\circ R}{860^\circ R} = 0.174$$

THE CORRECT ANSWER IS: (A)

129. $h_{mixture} = \dfrac{1}{2}(h_{water} + h_{steam}) = \dfrac{1}{2}(28.08 + 1{,}335.2)\text{Btu/lbm} = 681.64\text{ Btu/lbm}$

$$681.64\text{ Btu/lbm} = 180.1\text{ Btu/lbm} + x\left(970.4\text{ Btu/lbm}\right) \Rightarrow x = 0.516$$

THE CORRECT ANSWER IS: (C)

130. From $PV = n\overline{R}T$, $P/T = n\overline{R}/V$

THE CORRECT ANSWER IS: (D)

131. Flue gas mass for each component is number of moles × molecular weight:

9 CO_2	$\Rightarrow (9)(44) =$	396
9 H_2O	$\Rightarrow (9)(18) =$	162
50.76 N_2	$\Rightarrow (50.76)(28) =$	1,421
	Total	1,979

$$\%\ H_2O = \frac{162}{1{,}979} = 8.2\%$$

THE CORRECT ANSWER IS: (B)

132. At initial conditions of 600°F and 300 psia:

$$h_i = 1{,}314.5 \text{ Btu/lbm} \qquad s_i = 1.6266 \text{ Btu/(lbm-°R)}$$

Expansion process is constant entropy

i.e., $s_i = s_f$

Determine quality of steam at final condition

$$s_f = s_{\text{liquid}} + (x)s_{\text{vapor}}$$
$$1.6266 = 0.2009 + (x)1.6852$$
$$x = 0.846$$

Find h_f^t, theoretical enthalpy

$$h_f = h_{\text{liquid}} + (x)h_{\text{vapor}}$$
$$h_f = 109.37 + (0.846)(1.013.1)$$
$$h_f = 966.48 \text{ Btu/lbm}$$

Adiabatic efficiency:

$$\eta_{\text{adiabatic}} = \frac{\left(h_i - h_f^a\right)}{\left(h_i - h_f^t\right)}$$

$$0.8 = \frac{\left(1{,}314.5 - h_f^a\right)}{\left(1{,}314.5 - 966.48\right)}$$

$$h_f^a = 1{,}036.1 \text{ Btu/lb}$$

THE CORRECT ANSWER IS: (C)

MECHANICAL MORNING SOLUTIONS

133.

$\Delta T_A = 70°F$

$\Delta T_B = 40°F$

$$\text{LMTD} = \frac{\Delta T_A - \Delta T_B}{\ln\left(\frac{\Delta T_A}{\Delta T_B}\right)}$$

$$\text{LMTD} = \frac{70 - 40}{\ln\left(\frac{70}{40}\right)} = 53.6°\text{F}$$

Alternate Solution:

$$\text{LMTD} = \left[\frac{(\Delta T_A)^{1/3} + (\Delta T_B)^{1/3}}{2}\right]^3$$

$$= \frac{70^{1/3} + 40^{1/3}}{2} = 53.6°\text{F}$$

THE CORRECT ANSWER IS: (C)

134. Using a psychrometric chart, find the intersection of the 95°F line and the 25% rh line. Follow the constant humidity ratio line at 0.0088 lb H_2O/lb air horizontally to saturation condition where the dew point temperature is 54°F.

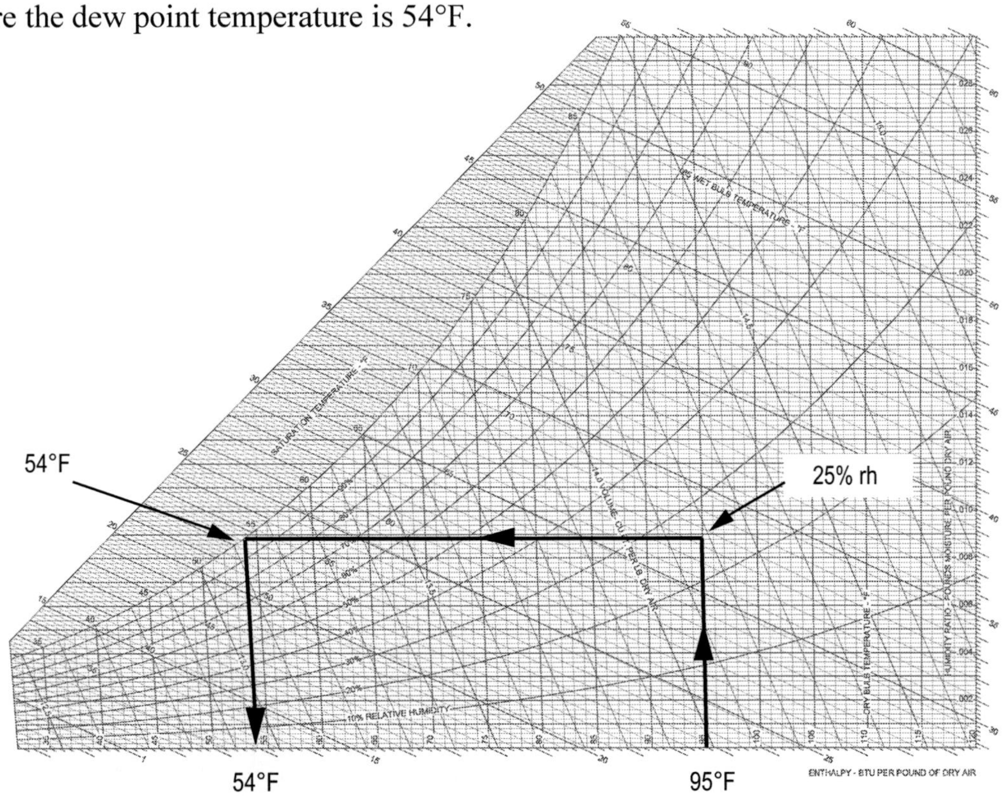

THE CORRECT ANSWER IS: (B)

135. Partial Pressure (p) = P_{SAT} at Temp × % rh

From steam table, P_{SAT} at 80°F = 0.5073 psi

p(80°F, 60% rh) = (0.5073) (0.6) = 0.30 psi

Alternate Solution:
Partial pressure = P_{SAT} at dew point
From psychrometric chart at 80°F, 60% rh – dew point temperature is 65°F
From steam table, P_{SAT} = 0.30 psi

THE CORRECT ANSWER IS: (B)

136. Throttling is a constant enthalpy process.
Using the T-s chart for R-134a, at 250 psi saturated liquid state, read $S_{initial}$ = 0.118 Btu/(lbm-°R).
Move down line of constant enthalpy to 40 psia – entropy, S_{final} is 0.126 Btu/(lbm-°R).
Change $\Delta S = S_{final} - S_{initial} = 0.126 - 0.118 = 0.008$

THE CORRECT ANSWER IS: (B)

137. An expansion valve would be used for isenthalpic expansion of high-pressure saturated liquid refrigerant on this, the throttling leg.

THE CORRECT ANSWER IS: (C)

138. Heat of vaporization of water = 970 Btu/lbm

Sensible heat load $Q_S = 3,000 \text{ W} \times \dfrac{3.41 \text{ Btu/hr}}{1 \text{ W}} = 10,230 \text{ Btu/hr}$

Latent heat load Q_l = 5 lbm/hr × 970 Btu/lbm = 4,850 Btu/hr

Total heat load $Q = Q_s + Q_l$ = 10,230 Btu/hr + 4,850 Btu/hr = 15,080 Btu/hr

THE CORRECT ANSWER IS: (D)

139. $$COP_{refrigerator} = \frac{Q_{absorbed}}{W_{compression}}$$

1 refrigeration ton = 200 Btu/min = 12,000 Btu/hr = 3,517 W

$$COP_{refrigerator} = \frac{3.517\ \text{kW}}{1\ \text{kW}} = 3.517$$

THE CORRECT ANSWER IS: (C)

140. Log mean temperature difference $$\text{LMTD} = \frac{\Delta T_A - \Delta T_B}{\ln\left(\frac{\Delta T_A}{\Delta T_B}\right)}$$

$$= \frac{(140°F - 75°F) - (140°F - 60°F)}{\ln\left(\frac{140°F - 75°F}{140°F - 60°F}\right)} = 72°F$$

Heat transfer $Q = UA(\text{LMTD}) = [8\ \text{Btu/(hr-ft}^2\text{-°F)}](20\ \text{ft}^2)(72°F) = 11{,}540\ \text{Btu/hr}$

$$Q = \dot{m}\Delta h \rightarrow \dot{m} = \frac{Q}{\Delta h} = \frac{Q}{h_{vapor} - h_{liquid}} = \frac{11{,}540\ \text{Btu/hr}}{120\ \text{Btu/lb} - 60\ \text{Btu/lb}} = 192\ \text{lbm/hr}$$

THE CORRECT ANSWER IS: (D)

THERMAL AND FLUIDS SYSTEMS
AFTERNOON SOLUTIONS

Answers to the THERMAL AND FLUIDS SYSTEMS Afternoon Sample Questions

Detailed solutions for each question begin on the next page.

501	D	**521**	C
502	B	**522**	B
503	A	**523**	B
504	D	**524**	B
505	D	**525**	B
506	C	**526**	B
507	C	**527**	C
508	A	**528**	B
509	B	**529**	D
510	B	**530**	A
511	D	**531**	D
512	B	**532**	A
513	B	**533**	C
514	B	**534**	C
515	C	**535**	D
516	B	**536**	A
517	C	**537**	D
518	A	**538**	C
519	C	**539**	C
520	C	**540**	D

THERMAL AND FLUIDS SYSTEMS AFTERNOON SOLUTIONS

501. $P = \rho g L$

$$P_{psig} = P_{psia} - P_{atm}$$
$$= 19.4 \text{ psia} - 14.7 \text{ psia}$$
$$= 4.7 \text{ psig}$$

$$\rho = \frac{P}{gL} = 4.7\frac{\text{lbf}}{\text{in}^2} \times 144\frac{\text{in}^2}{\text{ft}^2} \times \frac{\text{sec}^2}{32.2 \text{ ft}} \times \frac{1}{35 \text{ in.}} \times \frac{32.2 \text{ lbm-ft}}{\text{lbf-sec}^2} \times \frac{12 \text{ in.}}{\text{ft}}$$
$$= 232.0 \text{ lbm/ft}^3$$

THE CORRECT ANSWER IS: (D)

502. $m = \rho AV = AV/v$

From steam tables

At 2 psia: $v_L = 0.016230 \text{ ft}^3\text{/lbm}$

$v_V = 173.72 \text{ ft}^3\text{/lbm}$

X = steam quality = 1 – moisture content
= 1 – 0.07
= 0.93

$$v = v_L + X(v_V - v_L)$$
$$= 0.01623 + (0.93)(173.72 - 0.01623)$$
$$= 161.56 \text{ ft}^3\text{/lbm}$$

$m = AV/v$

900,000 lbm/hr = A (300 ft/sec)(3,600 sec/hr)/(161.56 ft^3/lbm)

$A = 135 \text{ ft}^2$

THE CORRECT ANSWER IS: (B)

503. V = Mach number × speed of sound

$$= 0.85 \times c$$
$$= (0.85)\,[kRT]^{0.5}$$
$$= (0.85)\left[(1.4)\left(287\ \frac{\text{J}}{\text{kg}\bullet\text{K}}\right)(-56.7°\text{C} + 273 \text{ K})\left(\frac{\text{N}\bullet\text{m}}{\text{J}}\right)\left(1\frac{\text{kg}\bullet\text{m}}{\text{s}^2\bullet\text{N}}\right)\right]^{0.5}$$

V = 251 m/s

THE CORRECT ANSWER IS: (A)

504. mRT = PAV

A = mRT/(PV)

= [(150 lbm/sec)(53.35 ft-lbf/lbm-°R)(430°R)]/[(5.45 lbf/in^2)(144 in^2/ ft^2)(600 ft/sec)] = 7.30 ft^2

THE CORRECT ANSWER IS: (D)

505. Pipe I.D. = 10.020 in.

Orifice diameter = (0.5)(10.020) = 5.010 in.

$$\text{Flow rate, } Q = C_f A_o \sqrt{\frac{2g\Delta P}{\rho}}$$

$$\frac{(1{,}200 \text{ gal/ min})}{(60 \text{ sec/min})(7.48 \text{ gal/ft}^3)} = (0.5)\left(\frac{\pi}{4}\right)(5.010 \text{ in.})^2(1 \text{ ft}/12 \text{ in.})^2\sqrt{\frac{(2)(32.174 \text{ lbm-ft/lbf-sec}^2)(\Delta P \text{ lbf/in}^2)}{(62 \text{ lbm/ft}^3)(1 \text{ ft}^2/144 \text{ in}^2)}}$$

$$\Delta P = 10.2 \text{ lbf/in}^2$$

THE CORRECT ANSWER IS: (D)

506. $$\text{Velocity} = \frac{\text{volumetric flowrate}}{\text{area}}$$

$$V = \frac{(300 \text{ gal/min})}{(7.48 \text{ gal/ft}^3)(60 \text{ sec/min})\left(\frac{\pi}{4}\right)(4.026 \text{ in.})^2\left(\frac{1 \text{ ft}}{12 \text{ in.}}\right)^2}$$

$$= 7.56 \text{ ft/sec}$$

$$h_f = \left[K_{ent} + K_{elbow} + f\frac{L_e}{D}\right]\frac{V^2}{2g}$$

$$= \left[0.5 + 0.9 + (0.018)\frac{(33.5 \text{ ft})}{(4.026 \text{ in.})(1 \text{ ft}/12 \text{ in.})}\right]\frac{(7.56 \text{ ft/sec})^2}{(2)(32.2 \text{ ft/sec}^2)}$$

$$h_f = 2.84 \text{ ft}$$

THE CORRECT ANSWER IS: (C)

507. $V = \dfrac{\dot{m}}{\rho A}$

$$= \frac{50 \text{ lbm/sec}}{\left(51.4 \text{ lbm/ft}^3\right)\left(\dfrac{\pi}{4}\right)\left(\dfrac{4.026 \text{ in.}}{12 \text{ in/ft}}\right)^2} \text{ ft-kips}$$

$$= 11.0 \text{ ft/sec}$$

$$Re = \frac{VD\rho}{\mu}$$

$$= \frac{(11.0 \text{ ft/sec})\left(\dfrac{4.026 \text{ in.}}{12 \text{ in./ft}}\right)\left(51.4 \text{ lbm/ft}^3\right)}{0.000156 \text{ lbm/(ft-sec)}}$$

$$Re = 1.216 \times 10^6$$

For liquid metals, including sodium

$$Nu = \frac{hD}{K} = 5.0 + 0.025(Re\ Pr)^{0.8}$$

From Incropera and Dewitt, *Fundamentals of Heat Transfer,* Pr = 0.0044

$$h = \frac{K}{D}\left[5.0 + (0.025)(Re\,Pr)^{0.8}\right]$$

$$= \frac{[37.6 \text{ Btu/(hr-ft-°F)}]}{(4.026 \text{ in./12 in./ft})}\left[5.0 + (0.025)\left(1.216 \times 10^6\right)^{0.8}(0.0044)^{0.8}\right]$$

$$h = 3{,}253 \text{ Btu/(hr-ft}^2\text{-°F)}$$

THE CORRECT ANSWER IS: (C)

508.

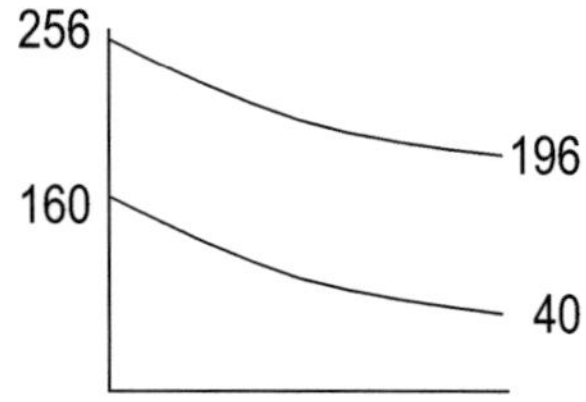

COUNTERFLOW

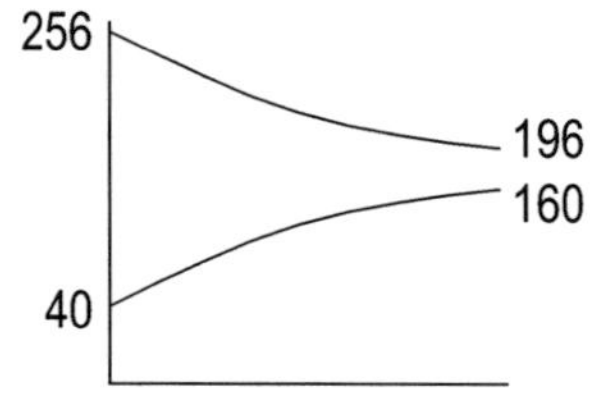

PARALLEL FLOW

$$\left(\dot{m}c_p\Delta T\right)_{oil} = \left(\dot{m}c_p\Delta T\right)_{water}$$

$$\dot{m}_{oil}c_{p_{oil}}\left(T_{oil\ in} - T_{oil\ out}\right) = \dot{m}_w c_{p\ w}\left(T_{w\ out} - T_{w\ in}\right)$$

$$(37{,}000\ \text{lbm/hr})\left[0.54\ \text{Btu/(lbm-°F)}\right](256\ \text{°F} - T_{oil\ out})$$

$$= (10{,}000\ \text{lbm/hr})\left[1\ \text{Btu/(lbm-°F)}\right](160\text{°F} - 40\text{°F})$$

$$T_{oil\ out} = 196\text{°F}$$

$$\text{LMTD} = \frac{\Delta T_A - \Delta T_B}{\ln\left(\dfrac{\Delta T_A}{\Delta T_B}\right)}$$

$\text{LMTD}_{\text{counterflow}}$: $\Delta T_A = (256 - 160) = 96\text{°F}$
$\Delta T_B = (196 - 40) = 156\text{°F}$

$\text{LMTD}_{\text{parallel flow}}$: $\Delta T_A = (256 - 40) = 216\text{°F}$
$\Delta T_B = (196 - 160) = 36\text{°F}$

$$\text{ratio} = \frac{\text{LMTD}_{CF}}{\text{LMTD}_{PF}}$$

$$= \frac{(96-156)/\ln(96/156)}{(216-36)/\ln(216/36)}$$

$$= 1.23$$

THE CORRECT ANSWER IS: (A)

509. When both radiation and convection occur, the heat transfer is the sum of the two. If the tube is small compared to the room:

$$A = \pi DL = \pi(0.5\ \text{ft})(20\ \text{ft}) = 31.42\ \text{ft}^2$$

$$Q = h\,A\,(T_s - T_a) + \sigma\,A\,\varepsilon\,[(T_s + 460)^4 - (T_a + 460)^4]$$

$$= [5\ \text{Btu/(hr-ft}^2\text{-°F)}](31.42\ \text{ft}^2)(450\text{°F} - 65\text{°F})$$

$$+ [0.1714 \times 10^{-8}\ \text{Btu/(hr-ft}^2\text{-°R}^4)](31.42\ \text{ft}^2)(0.5)[(450 + 460)^4 - (65 + 460)^4\text{°R}^4]$$

$$= 76{,}900\ \text{Btu/hr}$$

THE CORRECT ANSWER IS: (B)

510. $Q = \dfrac{A\Delta T}{R_{total}}$

$$\text{Without shade: } R_{total} = \frac{1}{h_{out}} + \frac{\Delta x}{K_{glass}} + \frac{1}{h_{in}}$$

$$= \frac{1}{4\text{ Btu/(hr-ft}^2\text{-°F)}} + \frac{(1/8\text{ in.})(1\text{ ft}/12\text{ in.})}{0.63\text{ Btu/(hr-ft-°F)}} + \frac{1}{1.46\text{ Btu/(hr-ft}^2\text{-°F)}}$$

$$= 0.9515\frac{\text{hr-ft}^2\text{-°F}}{\text{Btu}}$$

$$\text{With shade: } R_{total} = R_{without\ shade} + R_{air\ gap} + R_{shade}$$

$$= 0.9515\frac{\text{hr-ft}^2\text{-°F}}{\text{Btu}} + 0.90\frac{\text{hr-ft}^2\text{-°F}}{\text{Btu}} + \frac{(1/8\text{ in.})(1\text{ ft}/12\text{ in.})}{0.03\dfrac{\text{Btu}}{\text{hr-ft}^2\text{-°F}}}$$

$$= 2.1987\frac{\text{hr-ft}^2\text{-°F}}{\text{Btu}}$$

$$\Delta Q = Q_{without\ shade} - Q_{shade}$$

$$= \frac{(200\text{ ft}^2)(70\text{°F})}{0.9515\dfrac{\text{hr-ft}^2\text{-°F}}{\text{Btu}}} - \frac{(200\text{ ft}^2)(70\text{°F})}{2.1987\dfrac{\text{hr-ft}^2\text{-°F}}{\text{Btu}}}$$

$$= 8{,}346\text{ Btu/hr}$$

THE CORRECT ANSWER IS: (B)

511. $\text{Work} = \dot{m}c_p\Delta T$

$c_{p\ air} = 0.24\text{ Btu/(lbm-°R)}$

$$\text{Work} = [0.24\text{ Btu/(lbm-°R)}][(150\text{ lbm/sec} - 3\text{ lbm/sec})(900\text{°R} - 460\text{°R}) + (3\text{ lbm/sec})(630\text{°R} - 460\text{°R})]$$

$$= (15{,}645.6\text{ Btu/sec})\left(\frac{3{,}600\text{ sec}}{\text{hr}}\right)(3.929\times10^{-4}\text{ hp/Btu/sec})$$

$$= 22{,}130\text{ hp}$$

THE CORRECT ANSWER IS: (D)

512. For the throttling process (1) to (2)

$h_1 = h_2 = 1{,}224.6$ Btu/lbm (from steam table)

at State 3:

$$T_{sat} = 292.69°F$$
$$h_{sat} = 262.24 \text{ Btu/lbm}$$
$$h_3 = h_{sat} - c_p \Delta T$$
$$= 262.24 \text{ Btu/lbm} - [1 \text{ Btu/(lbm-°F)}](292.69°F - 102°F)$$
$$= 71.6 \text{ Btu/lbm}$$

at State 4:

$$T_{sat} = 292.69°F$$
$$h_{sat} = 1{,}177.8 \text{ Btu/lbm}$$
$$h_4 = h_{sat} - c_p \Delta T$$
$$= 1{,}177.8 \text{ Btu/lbm} + [0.45 \text{ Btu/(lbm-°R)}](7°R)$$
$$= 1{,}181.0 \text{ Btu/lbm}$$

$$\dot{m}_2 h_2 + \dot{m}_3 h_3 = \dot{m}_4 h_4$$

$$(100{,}000 \text{ lbm/hr})(1{,}224.6 \text{ Btu/lbm}) + \dot{m}_3(71.6 \text{ Btu/lbm}) = (100{,}000 \text{ lbm/hr} + \dot{m}_3)(1{,}181.0 \text{ Btu/lbm})$$

$$(100{,}000 \text{ lbm/hr})(1{,}224.6 \text{ Btu/lbm}) = -\dot{m}_3(71.6 \text{ Btu/lbm}) + \dot{m}_3(1{,}181.0 \text{ Btu/lbm}) + (100{,}000 \text{ lbm/hr})(1{,}181.0 \text{ Btu/lbm})$$

$$\dot{m}_3 = 3{,}930 \text{ lbm/hr}$$

THE CORRECT ANSWER IS: (B)

513. $$m_{fraction} = \frac{m_{He}}{m_{total}} = 0.09091 = \frac{m_{He}}{m_{He} + 15{,}000 \text{ lbm}}$$

$$m_{He} = 1{,}500 \text{ lbm}$$

THE CORRECT ANSWER IS: (B)

514. Feedwater in:

$\dot{m} = 300,000$ lbm/hr
$T = 460°F$
$c_p = 1$ Btu/(lbm-°F)

Feedwater out:

$\dot{m} = 300,000$ lbm/hr
$T = T_{in} + 80°F = 540°F$

Steam in:

$P = 1,000$ psia
$T = 650°F$
$h = 1,289.4$ Btu/lbm (from steam table)

Saturated liquid out:

$P = 1,000$ psia
$h = 542.65$ Btu/lbm

$$(\dot{m}\Delta h)_{feedwater} = (\dot{m}\Delta h)_{steam}$$

$$(300{,}000\text{ lbm/hr})[1\text{ Btu/(lbm-°F)}](80°F) = \dot{m}_{steam}\left(1{,}289.4\text{ Btu/lbm} - 542.65\text{ Btu/lbm}\right)$$

$$\dot{m}_{steam} = 32{,}139\text{ lbm/hr}$$

THE CORRECT ANSWER IS: (B)

515. Stoichiometric mass of air

= (13.5 mole O_2 × 32 lbm/mole O_2) + (50.8 mole N_2 × 28 lbm/mole N_2)

[illegible]4 lbm air

[illegible] per mole of fuel)

[illegible] × 1 lbm H)

[illegible] lbm air)/126 lbm fuel

[illegible]el

[illegible]R IS: (C)

516. (1) $P_1 = 900$ psia
$T_1 = 900°F$
From superheated steam properties
$h_1 = 1{,}452.3$ Btu/lbm
$s_1 = 1.6262$ Btu/(lbm-°R)

(2) $P_2 = 96$ psia
$T_2 = 450°F$
From superheated steam properties
$h_2 = 1{,}254.38$ Btu/lbm

For isentropic expansion to 96 psia
$s_2' = 1.6262$ Btu/(lbm-°R)
$h_2' = 1{,}201.24$ Btu/lbm

efficiency, $\eta = (h_1 - h_2)/(h_1 - h_2')$

$$= \frac{1{,}452.3 \text{ Btu/lbm} - 1{,}254.38 \text{ Btu/lbm}}{1{,}452.3 \text{ Btu/lbm} - 1{,}201.24 \text{ Btu/lbm}}$$

$= 0.79$ or 79%

THE CORRECT ANSWER IS: (B)

517. $q = \dot{m}_{air}(h_3 - h_2) = \dot{m}_{air}(c_p)(T_3 - T_2)$

For an isentropic, closed system

$$T_2 = T_1\left(\frac{P_2}{P_1}\right)^{(k-1)/k}$$

For air, $k = 1.4$ and $c_p = 0.24$ Btu/(lbm-°F)

$$T_2 = (80 + 460°R)\left(\frac{95.6 \text{ psia}}{14.7 \text{ psia}}\right)^{(1.4-1)/1.4}$$

$= 922.0°R$
$= 462°F$

$$q = (1 \text{ lbm}_{air})\left(0.24\frac{\text{Btu}}{\text{lbm-°F}}\right)(1{,}550°F - 462°F)$$

$= 261$ Btu/lbm

THE CORRECT ANSWER IS: (C)

518. From the psychrometric chart:

State 1 58°F db, 90% rh

$w_1 = 0.0092$ lbm vapor/lbm dry air

$v_1 = 13.25$ ft^3/lbm dry air

State 2 95°F db, 47% rh

$w_2 = 0.0167$ lbm vapor/lbm dry air

$v_2 = 14.35$ ft^3/lbm dry air

$$\frac{\dot{m}_{vapor}}{\dot{m}_{dry\ air}} = \left[\frac{\left(\frac{w_1 Q_1}{v_1}\right) + \left(\frac{w_2 Q_2}{v_2}\right)}{\left(\frac{Q_1}{v_1}\right) + \left(\frac{Q_2}{v_2}\right)}\right]$$

$$= \frac{\frac{(0.0092 \text{ lbm vapor/lbm dry air})(10{,}000 \text{ ft}^3\text{/min})}{(13.25 \text{ ft}^3\text{/lbm dry air})} + \frac{(0.0167 \text{ lbm vapor/lbm dry air})(2{,}000 \text{ ft}^3\text{/min})}{(14.35 \text{ ft}^3\text{/lbm dry air})}}{\frac{(10{,}000 \text{ ft}^3\text{/min})}{(13.25 \text{ ft}^3\text{/lbm dry air})} + \frac{(2{,}000 \text{ ft}^3\text{/min})}{(14.35 \text{ ft}^3\text{/lbm dry air})}}$$

$$\frac{\dot{m}_{vapor}}{\dot{m}_{dry\ air}} = 0.01037 \text{ lbm vapor/lbm dry air}$$

THE CORRECT ANSWER IS: (A)

519. Using the pump affinity laws:

$$\frac{h_2}{h_1} = \left(\frac{n_2}{n_1}\right)^2$$

$h_2 = (60 \text{ ft})(1{,}000 \text{ rpm}/1{,}750 \text{ rpm})^2$

$h_2 = 19.60$ ft

THE CORRECT ANSWER IS: (C)

520. Shaft horsepower = W_{hp}/η

$$W_{hp} = \dot{m}w$$
$$= \dot{m}v\Delta P$$
$$\dot{m} = \rho\dot{V}$$
$$v = 1/\rho$$
$$\text{bhp} = \rho\dot{V}(1/\rho)\Delta P$$
$$= \dot{V}\Delta P$$
$$= (25\text{ gal/min})(50\text{ lbf/in}^2)(\text{ft}^3/7.48\text{ gal})(\text{min}/60\text{ sec})(144\text{ in}^2/1\text{ ft}^2)(\text{hp-sec}/550\text{ lbf-ft})$$
$$W_{hp} = 0.73\text{ hp}$$
$$\text{shaft hp} = 0.73\text{ hp}/0.7$$
$$= 1.04\text{ hp}$$

THE CORRECT ANSWER IS: (C)

521. $\eta = \dfrac{h_1 - h_2}{h_1 - h_{2,s}} = \dfrac{w_t}{w_{t,s}}$

From steam tables:

h_1 (T = 900°F, P = 1,000 psia) = 1,448.1 Btu/lbm

s_1 = 1.612 Btu/(lbm-°R)

For isentropic process:

$s_2 = s_1$ = 1.612 Btu/(lbm-°R)

At P_2 = 2 psia

$$x_2 = \frac{s_2 - s_f}{s_{fg}}$$
$$= \frac{1.612 - 0.175}{1.7448}$$
$$= 82.17\%$$
$$h_{2,s} = h_f + x_2 h_{fg}$$
$$= 94.02\text{ Btu/lbm} + 0.8217 \times 1{,}022.1\text{ Btu/lbm}$$
$$= 933.89\text{ Btu/lbm}$$
$$w_t = \eta\left(h_1 - h_{2,s}\right)$$
$$= 0.8\left(1{,}448.1 - 933.89\right)\text{Btu/lbm}$$
$$= 411.4\text{ Btu/lbm}$$

THE CORRECT ANSWER IS: (C)

522. Compressor:

$$\frac{T_{2,s}}{T_1} = \left(\frac{P_2}{P_1}\right)^{(k-1)/k}$$

$$T_{2,s} = T_1\left(\frac{P_2}{P_1}\right)^{(k-1)/k}$$

For air: k = 1.4

$$= (80°\text{F} + 460)(10)^{0.4/1.4}$$

$$T_{2,s} = 1{,}043°\text{R}$$

$$W_{comp} = \frac{\dot{m}c_p(T_{2,s} - T_1)}{\eta}$$

c_p = 0.24 Btu/(lbm-°R)

$$W_{comp} = \left[\left(100\frac{\text{lbm}}{\text{sec}}\right)\left(0.24\frac{\text{Btu}}{\text{lbm-°R}}\right)(1{,}043 - 540)°\text{R}\right]\left(\frac{60\text{ sec}}{\text{min}}\right)\left(\frac{1}{0.85}\right)\left(\frac{\text{hp-min}}{42.44\text{ Btu}}\right)$$

$$W_{comp} = 20{,}062\text{ hp}$$

Fan:

$$T_{2,s} = (540°\text{R})(1.3)^{0.4/1.4}$$

$$= 582.0°\text{R}$$

$$W_{fan} = \left(400\frac{\text{lbm}}{\text{sec}}\right)\left(0.24\frac{\text{Btu}}{\text{lbm-°R}}\right)(582.0 - 540)°\text{R}\left(\frac{60\text{ sec}}{\text{min}}\right)\left(\frac{1}{0.85}\right)\left(\frac{\text{hp-min}}{42.44\text{ Btu}}\right)$$

$$= 6{,}711.78\text{ hp}$$

$W_{total} = W_{comp} + W_{fan}$ = 20,062 hp + 6,712 hp = 26,774 hp

THE CORRECT ANSWER IS: (B)

523. Energy balance:

$$[(\dot{m})(\text{HHV})]_{refuse} + [(\dot{m})(\text{HHV})]_{coal} = \dot{Q}_{fuel}$$

$$\eta\dot{Q}_{fuel} = \dot{m}(h_2 - h_1)$$

$$= 90{,}000\text{ lbm/hr}(1{,}452 - 280)\text{ Btu/lbm} = 105.48\times10^6\text{ Btu/hr}$$

$$\dot{Q}_{fuel} = \frac{105.48\times10^6\text{ Btu/hr}}{0.75} = 140.64\times10^6\text{ Btu/hr}$$

$$(4{,}600\text{ lbm/hr})(5{,}500\text{ Btu/lbm}) + (\dot{m}_{coal})(10{,}800\text{ Btu/lbm}) = 140.64\times10^6\text{ Btu/hr}$$

$$\dot{m}_{coal} = 10{,}680\text{ lbm/hr}$$

THE CORRECT ANSWER IS: (B)

524. $\dot{m}_f = (BSFC)(BHP)$

$$= \frac{0.65 \text{ lbm}}{\text{hp-hr}} \times 200 \text{ hp} \times \left(\frac{1 \text{ hr}}{60 \text{ min}}\right)$$

$$= 2.1667 \text{ lbm/min}$$

$$\rho = \frac{\dot{m}}{\dot{V}}$$

$$\dot{V} = \frac{\dot{m}}{\rho}$$

$$= 2.1667 \text{ lbm/min} \times \frac{\text{ft}^3}{42.8 \text{ lbm}} \times \frac{7.48 \text{ gal}}{\text{ft}^3}$$

$$\dot{V} = 0.3787 \text{ gpm}$$

THE CORRECT ANSWER IS: (B)

525. From the isentropic relationships:

$$T_{2s} = T_1\left(\frac{P_2}{P_1}\right)^{(k-1)/k}$$

$$= 760°R\left(\frac{116 \text{ psia}}{29 \text{ psia}}\right)^{\frac{0.285}{1.285}} = 1{,}034°R$$

$$\eta_c = \frac{T_{2s} - T_1}{T_2 - T_1}$$

$$\therefore T_2 = 760°R + \left(\frac{1{,}034°R - 760°R}{0.8}\right) = 1{,}102°R$$

$$T_{5s} = T_4\left(\frac{P_5}{P_4}\right)^{(k-1)/k}$$

$$= 2{,}360°R\left(\frac{29 \text{ psia}}{116 \text{ psia}}\right)^{\frac{0.285}{1.285}} = 1{,}735°R$$

$$\eta_t = \frac{T_5 - T_4}{T_{5s} - T_4}$$

$$\therefore T_5 = 2{,}360°R + 0.85(1{,}735°R - 2{,}360°R) = 1{,}829°R$$

$$\eta_{reg} = \frac{T_3 - T_2}{T_5 - T_2}$$

$$T_3 = T_2 + \eta_{reg}(T_5 - T_2)$$

$$= 1{,}102°R + 0.8(1{,}829°R - 1{,}102°R) = 1{,}684°R$$

$$Q_{in} = c_p\,(T_4 - T_3)$$

$$= [0.203 \text{ Btu/(lbm-°R)}](2{,}360 - 1{,}684)°R$$

$$= 137.3 \text{ Btu/lbm}$$

THE CORRECT ANSWER IS: (B)

526. Since one pump can deliver 150 gpm at 60 ft head, three pumps in parallel are required to deliver 400 gpm.

THE CORRECT ANSWER IS: (B)

527. $Pv = RT$

$$R = \frac{\overline{R}}{M}$$

$$= 1{,}545 \frac{\text{lbf-ft}}{\text{lb mol-°R}} \frac{\text{lb mol}}{29\ \text{lbm}}$$

$$= 53.27 \frac{\text{lbf-ft}}{\text{lbm-°R}}$$

$$v = \frac{\dot{V}}{\dot{m}}$$

$$\therefore \dot{V} = \frac{\dot{m}RT}{P}$$

$$= 3{,}100 \frac{\text{lbm}}{\text{sec}} \left(53.27 \frac{\text{lbf-ft}}{\text{lbm-°R}}\right) (1{,}036 + 460)\ \text{°R} \left(\frac{\text{in}^2}{14.9\ \text{lbf}}\right) \left(\frac{1\ \text{ft}^2}{144\ \text{in}^2}\right) \left(\frac{60\ \text{sec}}{1\ \text{min}}\right)$$

$$\dot{V} = 6.91 \times 10^6\ \text{cfm}$$

THE CORRECT ANSWER IS: (C)

528. By inspection, after plotting curve of the three pumps in parallel, intersection with system curve is at 65 ft.

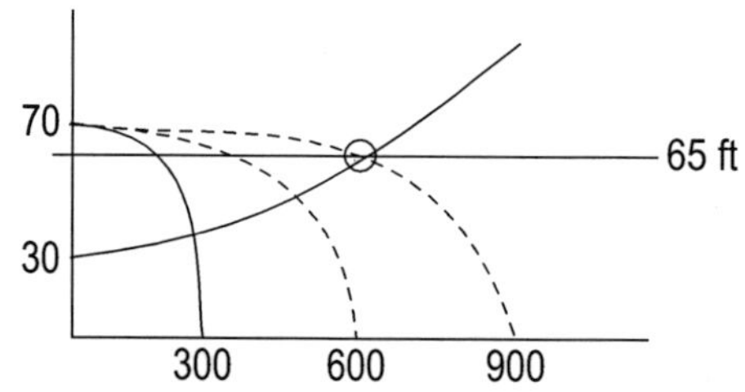

THE CORRECT ANSWER IS: (B)

529. $$NPSH = h_{atm} + h_z - h_f - h_{vp}$$

$$= \frac{P_a}{\rho g} + h_z - h_f - h_{vp}$$

$$= \left(\frac{10\ \text{lbf}}{\text{in}^2}\right) \left(\frac{\text{ft}^3}{60.3\ \text{lbm}}\right) \left(\frac{\text{sec}^2}{32.2\ \text{ft}}\right) \left(32.2 \frac{\text{lbm - ft}}{\text{lbf - sec}^2}\right) \left(\frac{144\ \text{in}^2}{1\ \text{ft}^2}\right) + (18 - 1.5)\ \text{ft} - 2.84\ \text{ft} - 23.06\ \text{ft}$$

$$= 14.5\ \text{ft}$$

THE CORRECT ANSWER IS: (D)

530. $$\frac{P_1}{\rho_1 g}+Z_1+\frac{V_1^2}{2g}+h=\frac{P_2}{\rho_2 g}+Z_2+\frac{V_2^2}{2g}+h_f$$

$$\rho_1=\rho_2=\rho$$

$$\therefore h=\frac{P_2-P_1}{\rho g}+\left(Z_2-Z_1\right)+\frac{V_2^2-V_1^2}{2g}+h_f$$

$$h_f=\left(K_{ent}+K_{elbow}+\frac{fL}{D}\right)\frac{V_1^2}{2g}$$

$$=\left(0.5+0.9+\frac{(0.018)(13.5+20)\text{ft}}{4.026\text{ in.}}\times\frac{12\text{ in}}{1\text{ ft}}\right)\times\frac{(7.6\text{ ft/sec})^2}{2\left(32.2\text{ ft/sec}^2\right)}$$

$$h_f=2.87\text{ ft}$$

With $V_1=0$

$$h=(120+14.7-10)\left(\frac{\text{lbf}}{\text{in}^2}\right)\left(\frac{144\text{ in}^2}{1\text{ ft}^2}\right)\left(\frac{\text{ft}^3}{60.3\text{ lbm}}\right)\left(\frac{32.2\text{ lbm-ft}}{\text{lbf-sec}^2}\times\frac{\text{sec}^2}{32.2\text{ ft}}\right)+(2.5-18)\text{ ft}$$

$$+\left(\frac{13^2\text{ ft}^2}{\text{sec}^2}\times\frac{\text{sec}^2}{2\times 32.2\text{ ft}}-0\right)+2.87\text{ ft}$$

$$h=287.8\text{ ft}$$

THE CORRECT ANSWER IS: (A)

$\frac{P_1}{\rho g}+z_1+\frac{V_1^2}{2g}+h=\frac{P_2}{\rho_2 g}+z_2+\frac{V_2^2}{2g}+h_f$

$\frac{(14.7+10)(144)}{60.3\cdot 3}+\frac{(7.6)^2}{2\cdot 32.2}+16.5+h=\frac{120(144)}{60.3\times 32.2}+\frac{13^2}{2\cdot 32.2}+0+2.8$

0.74 58.98 23.88 0.89 8.89 2.62 321.67

h = 267.28 ?

$V_1=0$

286.78

531. Energy balance

$$(\dot{m}c_pT)_{mixed} = (\dot{m}c_pT)_{RA} + (\dot{m}c_pT)_{OA}$$ where OA is Outside Air and RA is Return Air.

For the temperature range of interest the density is assumed constant.

$$(\rho VT)_{mixed} = (\rho VT)_{RA} + (\rho VT)_{OA}$$

$$T_{mixed} = \frac{(VT)_{RA} + (VT)_{OA}}{V_{mixed}}$$

$$= \frac{8{,}500 \text{ cfm} \times 75°\text{F} + 3{,}000 \text{ cfm} \times 40°\text{F}}{11{,}500 \text{ cfm}}$$

$$T_{mixed} = 65.9°\text{F}$$

Alternate solution:

Mix 3,000 cfm of outside air (40°F and 60% rh) with 8,500 cfm of return air (75°F and 50% rh) to produce 11,500 cfm supply air.

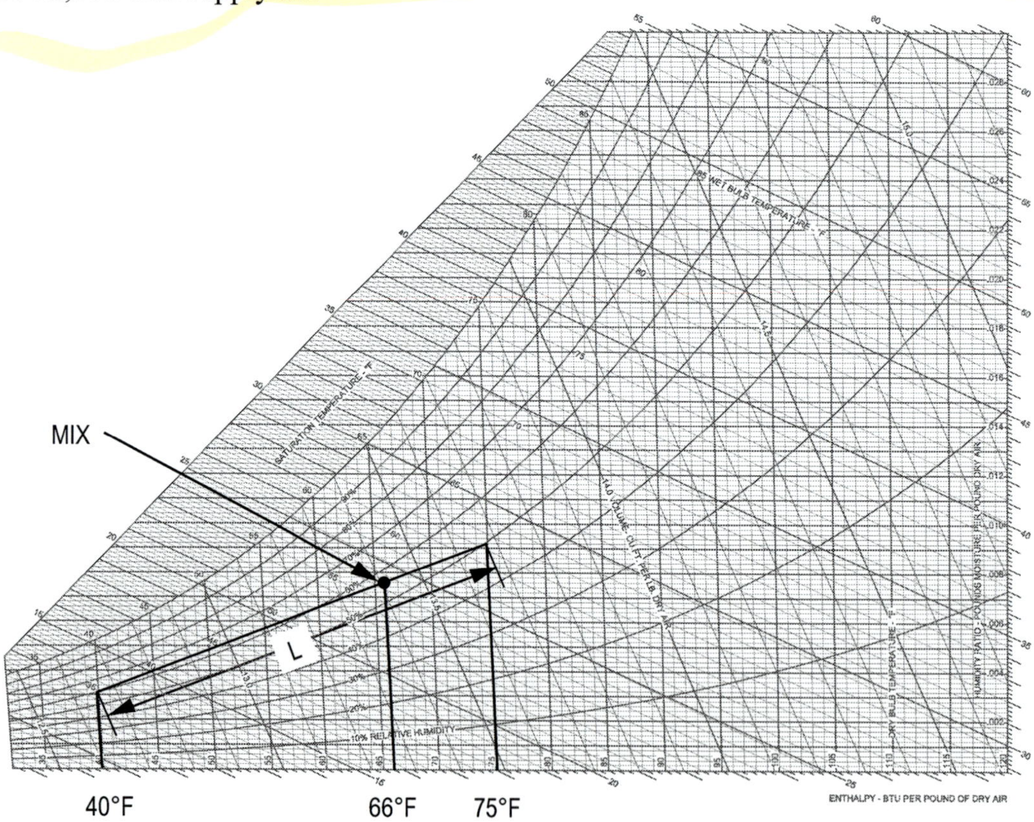

The mix is located at 3,000 cfm (L) /11,500 cfm = 0.261 (L) distance away from the room (return) air point along the line between the two states on the psychrometric chart. The dry bulb temperature of the mix is 66°F.

THE CORRECT ANSWER IS: (D)

532. P_C = pressure at caliper

P_{MC} = pressure at master cylinder

A_C = cross-sectional area of caliper

A_{MC} = cross-sectional area of master cylinder

F_{MC} = force at master cylinder

R_C = reaction at caliper cylinder

$$A_{MC} = \pi r_{MC}^2 = (\pi)(10\text{ mm})^2 = \pi(100\text{ mm}^2) = 314\text{ mm}^2$$

$$A_C = \pi r_C^2 = (\pi)(15\text{ mm})^2 = \pi(225\text{ mm}^2) = 707\text{ mm}^2$$

Hydraulic pressure is equal throughout the system. Therefore $P_{MC} = P_C$ where $P = \frac{F}{A}$.

The forces at the two cylinders are proportional to the cross-sectional areas.

$$\frac{F_{MC}}{A_{MC}} = \frac{R_C}{A_C}$$

$$F_{MC} = \frac{R_C}{A_C} A_{MC} = \frac{(200\text{ N})(314\text{ mm}^2)}{(707\text{ mm}^2)} = 88\text{ N}$$

THE CORRECT ANSWER IS: (A)

533. Consider the feedwater heater.

$$h_a\left(T = 500°F, P = 300\text{ psia}\right) = 1{,}257.5\text{ Btu / lbm}$$
$$h_b = h_f\left(P = 300\text{ psia}\right) = 394.1\text{ Btu / lbm}$$
$$h_c = 378.3\text{ Btu / lbm}$$
$$h_d = 85.8\text{ Btu / lbm}$$

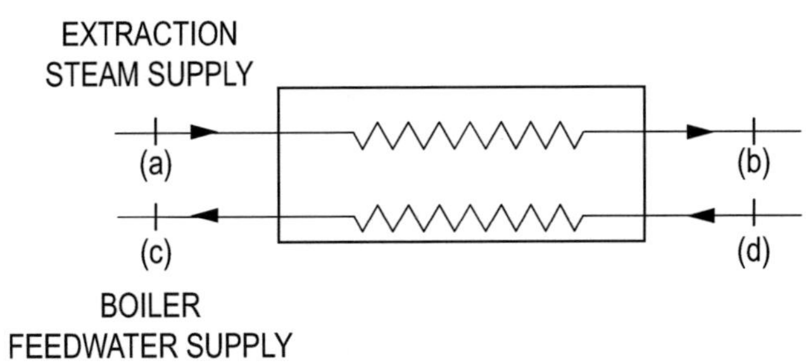

Energy balance:

$$m_{steam}\left(h_a - h_b\right) = m_{feedwater}\left(h_c - h_d\right)$$

$$\frac{m_{steam}}{m_{feedwater}} = \frac{\left(h_c - h_d\right)}{\left(h_a - h_b\right)} = \frac{378.3 - 85.8}{1{,}257.5 - 394.1} = 0.34 \frac{\text{lbm of steam}}{\text{lbm of feedwater}}$$

THE CORRECT ANSWER IS: (C)

534. $Q = AV$

$$A = \frac{\pi D^2}{4} = \left(\frac{\pi}{4}\right)(12\text{ in.})^2\left(\frac{1\text{ ft}}{12\text{ in.}}\right)^2 = 0.7854\text{ ft}^2$$

$$V = \sqrt{2gh}$$

$$h = (3\text{ in. H}_2\text{O})\left(\frac{1\text{ ft}}{12\text{ in.}}\right)\frac{\left(62.4\text{ lbm water/ft}^3\right)}{\left(0.075\text{ lbm air/ft}^3\right)} = 208\text{ ft air}$$

$$Q = \left(0.7854\text{ ft}^2\right)\left(\sqrt{(2)(32.2\text{ ft/sec}^2)(208\text{ ft})}\right)\left(\frac{60\text{ sec}}{1\text{ min}}\right)$$

$$Q = 5{,}454\text{ ft}^3\text{/min}$$

THE CORRECT ANSWER IS: (C)

535. Fuel use increases in order to add additional energy to the steam. Therefore, CO_2 emissions will increase.

THE CORRECT ANSWER IS: (D)

536. Extend line B until it intersects with saturated line. The temperature is the dew point of 50°F.

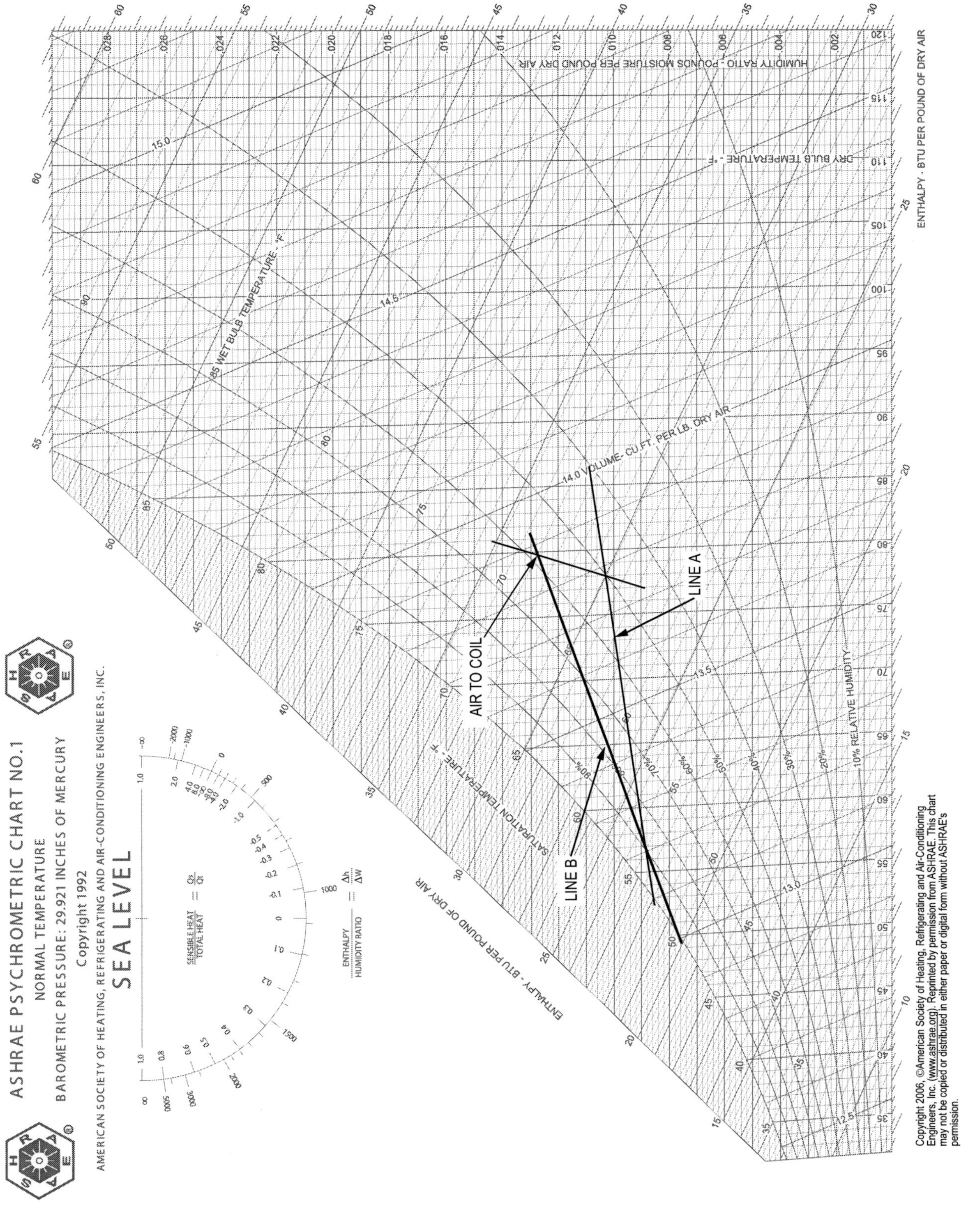

THE CORRECT ANSWER IS: (A)

537. Evaluate the reduction in ash disposal costs ($/ton of displaced gob coal) by using pellets.

Ash costs – basis is 1 ton of gob coal:

Coal: 1.0 ton coal × 0.40 wt fraction ash × $10/ton ash = $4.00

Pellets: 1.0 ton coal × (HV ratio 5,500/6,550) × 0.10 wt fraction ash × $10/ton ash = $0.84

ΔCost = 4.00 – 0.84 = $3.16

THE CORRECT ANSWER IS: (D)

538. All carbon is converted to CO_2

$$C \rightarrow CO_2$$

Mass of coal/day = power × heat rate × HHV

$$= (400 \times 10^3 \text{ kW})\left(11{,}010 \frac{\text{Btu}}{\text{kWh}} \times \frac{24 \text{ hr}}{\text{day}}\right)\left(\frac{\text{lbm}}{14{,}100 \text{ Btu}}\right)$$

$$= 7.5 \times 10^6 \frac{\text{lbm}}{\text{day}}$$

Mass of CO_2/day = mass of coal/day × % of carbon × $\dfrac{\text{MW of } CO_2}{\text{MW of C}}$

$$= 7.5 \times 10^6 \frac{\text{lbm}}{\text{day}} \times 0.805 \times \frac{44}{12}$$

$$= 22.13 \times 10^6 \text{ lbm } CO_2\text{/day}$$

THE CORRECT ANSWER IS: (C)

539. ASME

THE CORRECT ANSWER IS: (C)

540. According to paragraph 7.1 of ASTM 372-1999, forging shall be subject to magnetic particle examination.

THE CORRECT ANSWER IS: (D)

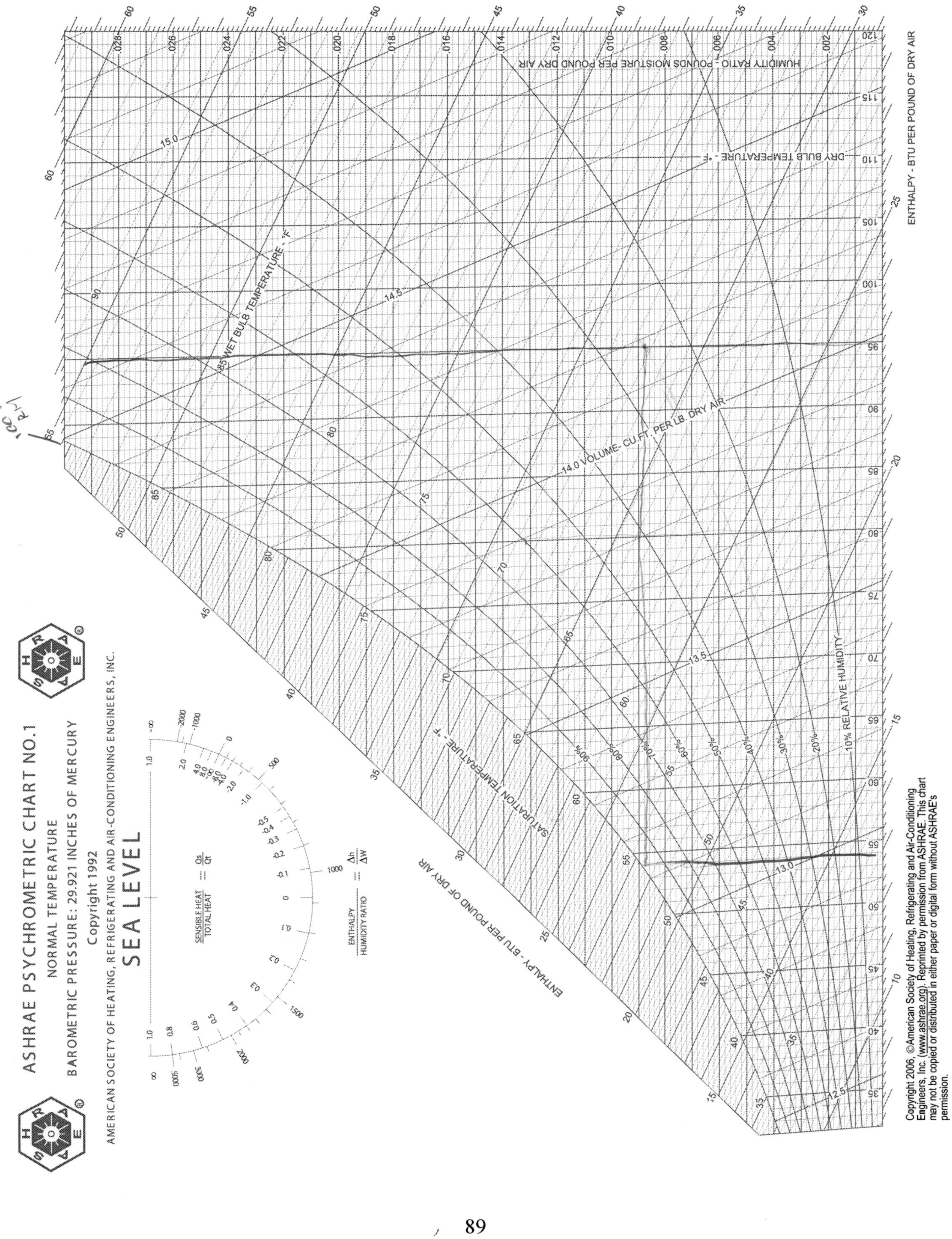
ASHRAE PSYCHROMETRIC CHART NO.1
NORMAL TEMPERATURE
BAROMETRIC PRESSURE: 29.921 INCHES OF MERCURY
Copyright 1992
AMERICAN SOCIETY OF HEATING, REFRIGERATING AND AIR-CONDITIONING ENGINEERS, INC.
SEA LEVEL
SENSIBLE HEAT / TOTAL HEAT = Qs / Qt
ENTHALPY / HUMIDITY RATIO = Δh / ΔW
WET BULB TEMPERATURE - °F
SATURATION TEMPERATURE - °F
14.0 VOLUME- CU.FT. PER LB. DRY AIR
10% RELATIVE HUMIDITY
ENTHALPY - BTU PER POUND OF DRY AIR
HUMIDITY RATIO - POUNDS MOISTURE PER POUND DRY AIR
DRY BULB TEMPERATURE - °F
ENTHALPY - BTU PER POUND OF DRY AIR
Copyright 2006, ©American Society of Heating, Refrigerating and Air-Conditioning Engineers, Inc. (www.ashrae.org). Reprinted by permission from ASHRAE. This chart may not be copied or distributed in either paper or digital form without ASHRAE's permission.

PE Sample Questions and Solutions Published by NCEES

Chemical
Civil: Construction
Civil: Geotechnical
Civil: Structural
Civil: Transportation
Civil: Water Resources and Environmental
Electrical and Computer: Computer Engineering
Electrical and Computer: Electrical and Electronics
Electrical and Computer: Power
Environmental
Mechanical: HVAC and Refrigeration
Mechanical: Mechanical Systems and Materials
Structural